NANOTECHNOLOGY
A Gentle Introduction to the Next Big Idea

Mark Ratner
Daniel Ratner

Prentice Hall
Professional Technical Reference
Upper Saddle River, NJ 07458
www.phptr.com

Library of Congress Cataloging-in-Publication Data

Ratner, Mark A.
Nanotechnology: a gentle introduction to the next big idea / Mark A. Ratner, Daniel Ratner.
p. cm.
ISBN 0-13-101400-5 (pbk.)
1. Nanotechnology. I. Ratner, Daniel. II. Title.

T174.7 .R38 2002
620'.5—dc21

2002035488

Editorial / production supervision: BooksCraft, Inc., Indianapolis, IN
Publisher: Bernard Goodwin
Editorial assistant: Michelle Vincenti
Marketing manager: Dan DePasquale
Manufacturing manager: Alexis Heydt-Long
Cover design director: Jerry Votta
Art director: Gail Cocker-Bogusz
Color insert designer: Meg VanArsdale
Full-service production manager: Anne Garcia

Publishing as Prentice Hall PTR
Upper Saddle River, New Jersey 07458

Prentice Hall books are widely used by corporations and government agencies for training, marketing, and resale.

For information regarding corporate and government bulk discounts please contact:
Corporate and Government Sales (800) 382-3419 or corpsales@pearsontechgroup.com

Printed in the United States of America
10 9 8 7 6 Sixth Printing

ISBN 0-13-101400-5

Pearson Education Ltd.
Pearson Education Australia PTY,. Limited
Pearson Education Singapore, Pte. Ltd.
Pearson Education North Asia Ltd.
Pearson Education Canada Ltd.
Pearson Educación de Mexico, S. A. de C. V.
Pearson Education—Japan
Pearson Education Malaysia, Pte. Ltd.

For Nancy with Love

With many thanks for the enthusiasm, advice, support, and editorial assistance from Genevieve and Stacy, without whom none of this would be here.

We would also like to thank the DoD and NSF for their funding and support for nanotechnology research.

About Prentice Hall Professional Technical Reference

With origins reaching back to the industry's first computer science publishing program in the 1960s, Prentice Hall Professional Technical Reference (PH PTR) has developed into the leading provider of technical books in the world today. Formally launched as its own imprint in 1986, our editors now publish over 200 books annually, authored by leaders in the fields of computing, engineering, and business.

Our roots are firmly planted in the soil that gave rise to the technological revolution. Our bookshelf contains many of the industry's computing and engineering classics: *Kernighan and Ritchie's C Programming Language, Nemeth's UNIX System Administration Handbook, Horstmann's Core Java,* and *Johnson's High-Speed Digital Design.*

PH PTR acknowledges its auspicious beginnings while it looks to the future for inspiration. We continue to evolve and break new ground in publishing by providing today's professionals with tomorrow's solutions.

PRENTICE HALL PTR

Contents

Preface

This book has a straightforward aim—to acquaint you with the whole idea of nanoscience and nanotechnology. This comprises the fabrication and understanding of matter at the ultimate scale at which nature designs: the molecular scale. Nanoscience occurs at the intersection of traditional science and engineering, quantum mechanics, and the most basic processes of life itself. Nanotechnology encompasses how we harness our knowledge of nanoscience to create materials, machines, and devices that will fundamentally change the way we live and work.

Nanoscience and nanotechnology are two of the hottest fields in science, business, and the news today. This book is intended to help you understand both of them. It should require the investment of about six hours—a slow Sunday afternoon or an airplane trip from Boston to Los Angeles. Along the way, we hope that you will enjoy this introductory tour of nanoscience and nanotechnology and what they might mean for our economy and for our lives.

The first two chapters are devoted to the big idea of nanoscience and nanotechnology, to definitions, and to promises. Chapters 3 and 4 discuss the science necessary to understand nanotechnology; you can skip these if you remember some of your high school science and mathematics. Chapter 5 is a quick grand tour of some of the thematic areas of nanotechnology, via visits to laboratories. Chapters 6 to 9 are the heart of the book. They deal with the topical areas in which nanoscience and nanotechnology are concentrated: smart materials, sensors, biological structures, electronics, and optics. Chapters 10 and 11 discuss business applications and the relationship of nanotechnology to individuals in the society. The book ends with lists of sources of additional information about nanotechnology, venture capitalists who have expressed interest in nanotechnology, and a glossary of key nanotechnology terms. If you want to discuss nanotechnology or find links to more resources, you can also visit the book's Web site at www.nanotechbook.com.

We are grateful to many colleagues for ideas, pictures, and inspiration, and to Nancy, Stacy, and Genevieve for their editing, encouragement, and support. Mark Ratner thanks his students from Ari to Emily, colleagues, referees, and funding agents (especially DoD and NSF) for allowing him to learn something about the nanoscale. Dan Ratner wishes to thank his coworkers, especially John and the Snapdragon crew, for being the best and strongest team imaginable, and Ray for his mentoring. Thanks also to Bernard, Anne, Don, Sara, and everyone from Prentice Hall for making it possible.

We enjoyed the writing and hope you enjoy the read.

1 Introducing Nano

Nanotechnology is truly a portal opening on a new world.

Rita Colwell
Director, National Science Foundation

In this chapter...

WHY DO I CARE ABOUT NANO? . . . .

Over the past few years, a little word with big potential has been rapidly insinuating itself into the world's consciousness. That word is "nano." It has conjured up speculation about a seismic shift in almost every aspect of science and engineering with implications for ethics, economics, international relations, day-to-day life, and even humanity's conception of its place in the universe. Visionaries tout it as the panacea for all our woes. Alarmists see it as the next step in biological and chemical warfare or, in extreme cases, as the opportunity for people to create the species that will ultimately replace humanity.

While some of these views are farfetched, nano seems to stir up popular, political, and media debate in the same way that space travel and the Internet did in their respective heydays. The federal government spent more than $422 million on nano research in 2001. In 2002, it is scheduled to spend more than $600 million on nano programs, even though the requested budget was only $519 million, making nano possibly the only federal program to be awarded more money than was requested during a period of general economic distress. Nano is also among the only growth sectors in federal spending not exclusively related to defense or counterterrorism, though it does have major implications for national security.

Federal money for nano comes from groups as diverse as the National Science Foundation, the Department of Justice, the National Institutes for Health, the Department of Defense, the Environmental Protection Agency, and an alphabet soup of other government agencies and departments. Nano's almost universal appeal is indicated by the fact that it has political support from both sides of the aisle—Senator Joseph Lieberman and former Speaker-of-the-House Newt Gingrich are two of nano's most vocal promoters, and the National Nanotechnology Initiative (NNI) is one of the few Clinton-era programs strongly backed by the Bush administration.

The U.S. government isn't the only organization making nano a priority. Dozens of major universities across the world—from Northwestern University in the United States to Delft University of Technology in the Netherlands and the National Nanoscience Center in Beijing, China—are building new faculties, facilities, and research groups for nano. Nano research also crosses scientific disciplines.

Chemists, biologists, doctors, physicists, engineers, and computer scientists are all intimately involved in nano development.

Nano is big business. The National Science Foundation predicts that nano-related goods and services could be a $1 trillion market by 2015, making it not only one of the fastest-growing industries in history but also larger than the combined telecommunications and information technology industries at the beginning of the technology boom in 1998. Nano is already a priority for technology companies like HP, NEC, and IBM, all of whom have developed massive research capabilities for studying and developing nano devices. Despite this impressive lineup, well-recognized abbreviations are not the only organizations that can play. A host of start-ups and smaller concerns are jumping into the nano game as well. Specialty venture capital funds, trade shows, and periodicals are emerging to support them. Industry experts predict that private equity spending on nano could be more than $1 billion in 2002. There is even a stock index of public companies working on nano.

In the media, nano has captured headlines at CNN, MSNBC, and almost every online technical, scientific, and medical journal. The Nobel Prize has been awarded several times for nano research, and the Feynman Prize was created to recognize the accomplishments of nanoscientists. *Science* magazine named a nano development as Breakthrough of the Year in 2001, and nano made the cover of *Forbes* the same year, subtitled "The Next Big Idea." Nano has hit the pages of such futurist publications as *Wired Magazine*, found its way into science fiction, and been the theme of episodes of *Star Trek: The Next Generation* and *The X-Files* as well as a one-liner in the movie *Spiderman*.

In the midst of all this buzz and activity, nano has moved from the world of the future to the world of the present. Innovations in nano-related fields have already sparked a flurry of commercial inventions from faster-burning rocket fuel additives to new cancer treatments and remarkably accurate and simple-to-use detectors for biotoxins such as anthrax. Nano skin creams and suntan lotions are already on the market, and nano-enhanced tennis balls that bounce longer appeared at the 2002 Davis Cup. To date, most companies that claim to be nano companies are engaging in research or trying to cash in on hype rather than working toward delivering a true nano product, but there certainly are exceptions. There is no shortage of opinions on where nano can go and what it can mean, but both pundits and critics agree on one point—no matter who you are and what your business and interests may be, this science and its spin-off technologies have the potential to affect you greatly.

There are also many rumors and misconceptions about nano. Nano isn't just about tiny little robots that may or may not take over the world. At its core, it is a great step forward for science. NNI is already calling it "The Next Industrial Revolution"—a phrase they have imprinted on a surface smaller than the width of a human hair in letters 50 nanometers wide. (See Figure 1.1.)

For the debate on nano to be a fruitful one, everyone must know a little bit about what nano is. This book will address that goal, survey the state of the art, and offer some thoughts as to where nano will head in the next few years.

WHO SHOULD READ THIS BOOK? . . . .

This book is designed to be an introduction to the exciting fields of nanotechnology and nanoscience for the nonscientist. It is aimed squarely at the professional reader who has been hearing the buzz about nano and wants to know what it's all about. It is chiefly concerned with the science, technology, implications, and future of nano, but some of the business and financial aspects are covered briefly as well. All the science required to understand the book is reviewed in Chapter 3. If you have taken a high school or college chemistry or physics class, you will be on familiar ground.

We have tried to keep the text short and to the point with references to external sources in case you want to dig deeper into the subjects that interest you most. We have also tried to provide the essential vocabulary to help you understand what you read in the media and trade press coverage of nano while keeping this text approachable and easy to read. We've highlighted key terms where they are first defined and included a glossary at the end.

We hope that this book will be a quick airplane or poolside read that will pique your interest in nano and allow you to discuss nano

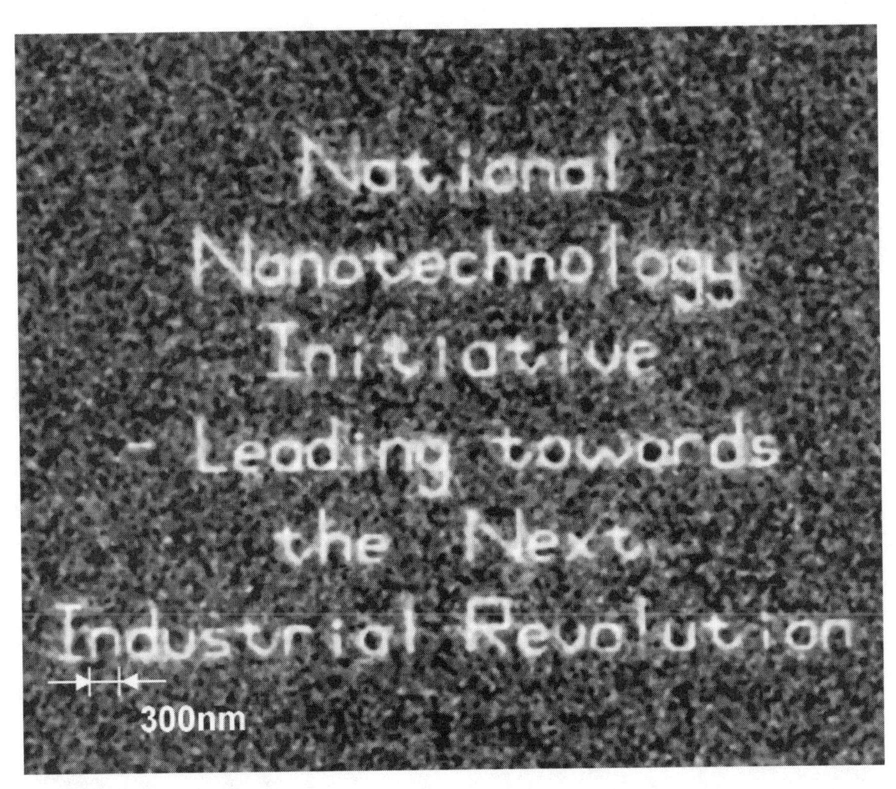

Figure 1.1
The Next Industrial Revolution, an image of a nanostructure. *Courtesy of the Mirkin Group, Northwestern University.*

with your friends and fascinate the guests at your next dinner party. Nano will be at the center of science, technology, and business for the next few years, so everyone should know a bit about it. We have designed this book to get you started. Enjoy!

WHAT IS NANO? A DEFINITION . . . .

When Neil Armstrong stepped onto the moon, he called it a small step for man and a giant leap for mankind. Nano may represent another giant leap for mankind, but with a step so small that it makes Neil Armstrong look the size of a solar system.

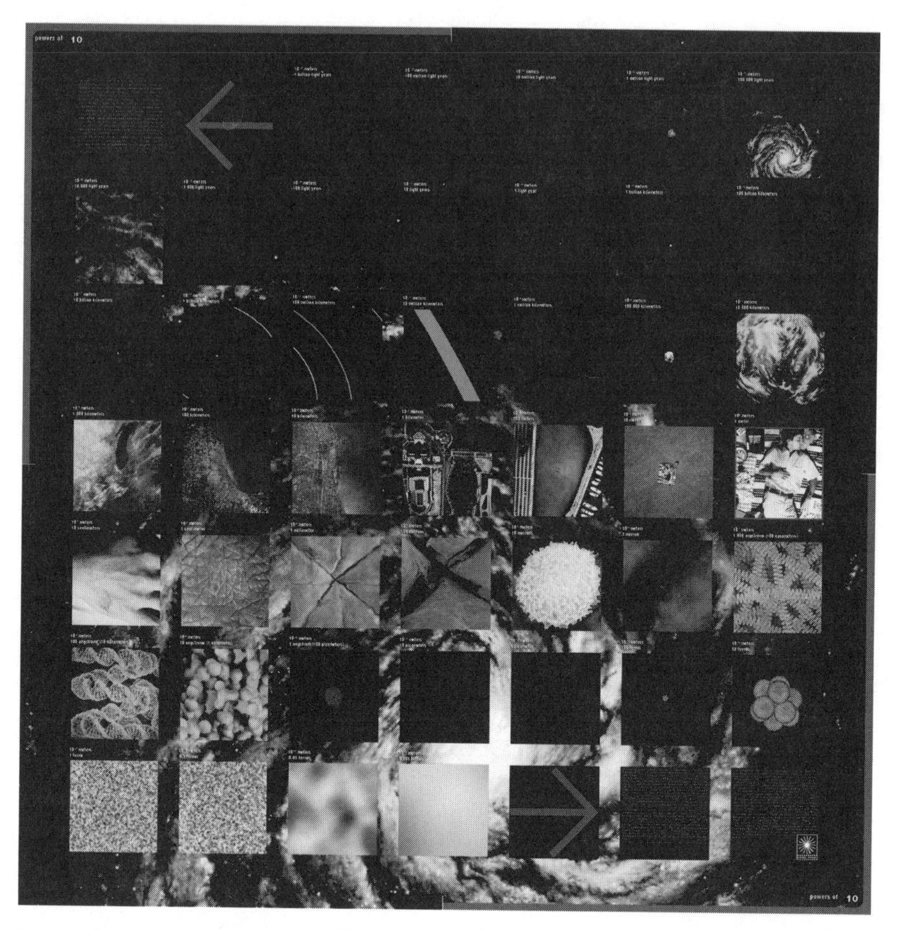

Figure 1.2
This image shows the size of the nanoscale relative to some things we are more familiar with. Each panel is magnified 10 times from the panel before it. As you can see, the size difference between a nanometer and a person is roughly the same as the size difference between a person and the orbit of the moon. *© 2001 Lucia Eames/Eames Office* (www.eamesoffice.com).

The prefix "nano" means one billionth. One nanometer (abbreviated as 1 nm) is 1/1,000,000,000 of a meter, which is close to 1/1,000,000,000 of a yard. To get a sense of the nano scale, a human hair measures 50,000 nanometers across, a bacterial cell measures a few hundred nanometers across, and the smallest features that are commonly etched on a commercial microchip as of February 2002 are around 130 nanometers across. The smallest things seeable with the unaided human eye are 10,000 nanometers across. Just ten hydrogen atoms in a line make up one nanometer. It's really very small indeed. See Figure 1.2.

Nanoscience is, at its simplest, the study of the fundamental principles of molecules and structures with at least one dimension roughly between 1 and 100 nanometers. These structures are known, perhaps uncreatively, as *nanostructures*. *Nanotechnology* is the application of these nanostructures into useful *nanoscale* devices. That isn't a very sexy or fulfilling definition, and it is certainly not one that seems to explain the hoopla. To explain that, it's important to understand that the nanoscale isn't just small, it's a special kind of small.

Anything smaller than a nanometer in size is just a loose atom or small molecule floating in space as a little dilute speck of vapor. So nanostructures aren't just smaller than anything we've made before, they are the smallest solid things it is possible to make. Additionally, the nanoscale is unique because it is the size scale where the familiar day-to-day properties of materials like conductivity, hardness, or melting point meet the more exotic properties of the atomic and molecular world such as wave-particle duality and quantum effects. At the nanoscale, the most fundamental properties of materials and machines depend on their size in a way they don't at any other scale. For example, a nanoscale wire or circuit component does not necessarily obey Ohm's law, the venerable equation that is the foundation of modern electronics. Ohm's law relates current, voltage, and resistance, but it depends on the concept of electrons flowing down a wire like water down a river, which they cannot do if a wire is just one atom wide and the electrons need to traverse it one by one. This coupling of size with the most fundamental chemical, electrical, and physical properties of materials is key to all nanoscience. A good and concise definition of nanoscience and nanotechnology that captures the special properties of the nanoscale comes from a National Science Foundation document edited by Mike Roco and issued in 2001:

> One nanometer (one billionth of a meter) is a magical point on the dimensional scale. Nanostructures are at the confluence of the smallest of human-made devices and the largest molecules of living things. Nanoscale science and engineering here refer to the fundamental understanding and resulting technological advances arising from the exploitation of new physical, chemical and biological properties of systems that are intermediate in size, between isolated atoms and molecules and bulk materials, where the transitional properties between the two limits can be controlled.

Figure 1.3
The nanoscale abacus. The individual bumps are molecules of carbon-60, which are about 1 nanometer wide. *Courtesy of J. Gimzewski, UCLA.*

Although both fields deal with very small things, nanotechnology should not be confused with its sister field, which is even more of a mouthful—microelectromechanical systems (*MEMS*). MEMS scientists and engineers are interested in very small robots with manipulator arms that can do things like flow through the bloodstream, deliver drugs, and repair tissue. These tiny robots could also have a host of other applications including manufacturing, assembling, and repairing larger systems. MEMS is already used in triggering mechanisms for automobile airbags as well as other applications. But while MEMS does have some crossover with nanotechnology, they are by no means the same. For one thing, MEMS is concerned with structures between 1,000 and 1,000,000 nanometers, much bigger than the nanoscale. See Figure 1.3. Further, nanoscience and nanotechnology are concerned with all properties of structures on the nanoscale, whether they are chemical, physical, quantum, or mechanical. It is more diverse and stretches into dozens of subfields. Nanotech is not nanobots.

In the next few chapters, we'll look in more depth at the "magical point on the dimensional scale," offer a quick recap of some of the basic science involved, and then do a grand tour of nanotech's many faces and possibilities.

A NOTE ON MEASURES

Almost all nanoscience is discussed using SI (mostly metric) measurement units. This may not be instinctive to readers brought up in the American system and not all the smaller measurements are frequently used. A quick list of small metric measures follows to help set the scale as we move forward into the world of the very small.

SI Unit (abbreviation)	Description
meter (m)	Approximately three feet or one yard
centimeter (cm)	1/100 of a meter, around half an inch
millimeter (mm)	1/1,000 of a meter
micrometer (μm)	1/1,000,000 of a meter; also called a micron, this is the scale of most integrated circuits and MEMS devices
nanometer (nm)	1/1,000,000,000 of a meter; the size scale of single small molecules and nanotechnology

2 Size Matters

In small proportions we just beauties see,
And in short measures life may perfect be.

Ben Jonson

In this chapter...

A DIFFERENT KIND OF SMALL . . . .

Imagine something we would all like to have: a cube of gold that is 3 feet on each side. Now take the imaginary cube and slice it in half along its length, width, and height to produce eight little cubes, each 18 inches (50 centimeters) on a side. The properties (excepting cash value) of each of the eight smaller cubes will be exactly the same as the properties of the big one: each will still be gold, yellow, shiny, and heavy. Each will still be a soft, electrically conductive metal with the same melting point it had before you cut it. Aside from making your gold a bit easier to carry, you won't have accomplished much at all.

Now imagine taking one of the eight 18-inch (50-centimeter) cubes and cutting it the same way. Each of the eight resulting cubes will now be 9 inches (25 centimeters) on a side and will have the same properties as the parent cube before we started cutting it. If we continue cutting the gold in this way and proceed down in size from feet to inches, from inches to centimeters, from centimeters to millimeters, and from millimeters to microns, we will still notice no change in the properties of the gold. Each time, the gold cubes will get smaller. Eventually we will not be able to see them with the naked eye and we'll start to need some fancy tools to keep cutting. Still, all the gold bricks' physical and chemical properties will be unchanged. This much is obvious from our real-world experience—at the macroscale chemical and physical properties of materials are not size dependent. It doesn't matter whether the cubes are gold, iron, lead, plastic, ice, or brass.

When we reach the nanoscale, though, everything will change, including the gold's color, melting point, and chemical properties. The reason for this change has to do with the nature of the interactions among the atoms that make up the gold, interactions that are averaged out of existence in the bulk material. Nano gold doesn't act like bulk gold.

The last few steps of the cutting required to get the gold cube down to the nanoscale represent a kind of *nanofabrication*, or nanoscale manufacturing. Starting with a suitcase-sized chunk of gold, our successive cutting has brought it down to the nanoscale. This particular kind of nanofabrication is sometimes called *top-down nanofabrication* because we started with a large structure and proceeded to make it smaller. Conversely, starting with individual atoms and building up to a nanostructure is called *bottom-up nanofabrication*. The tiny gold nanostructures that we prepared are sometimes called *quantum dots* or *nanodots* because they are roughly dot-shaped and have diameters at the nanoscale.

The process of nanofabrication, in particular the making of gold nanodots, is not new. Much of the color in the stained glass windows found in medieval and Victorian churches and some of the glazes found in ancient pottery depend on the fact that nanoscale properties of materials are different from macroscale properties. In particular, nanoscale gold particles can be orange, purple, red, or greenish, depending on their size. In some senses, the first nanotechnologists were actually glass workers in medieval forges (Figure 2.1) rather than the bunny-suited workers in a modern semiconductor plant

Figure 2.1
Early nanotechnologist. *Courtesy of Getty Images.*

Figure 2.2
Modern nanotechnologist. *Courtesy of Getty Images.*

(Figure 2.2). Clearly the glaziers did not understand why what they did to gold produced the colors it did, but now we do.

The size-dependent properties of the nanostructures cannot be sustained when we climb again to the macroscale. We can have a macroscopic spread of gold nanodots that looks red because of the size of the individual nanodots, but the nanodots will rapidly start looking yellow again if we start pushing them back together and let them join. Fortunately, if enough of the nanodots are close to each other but not

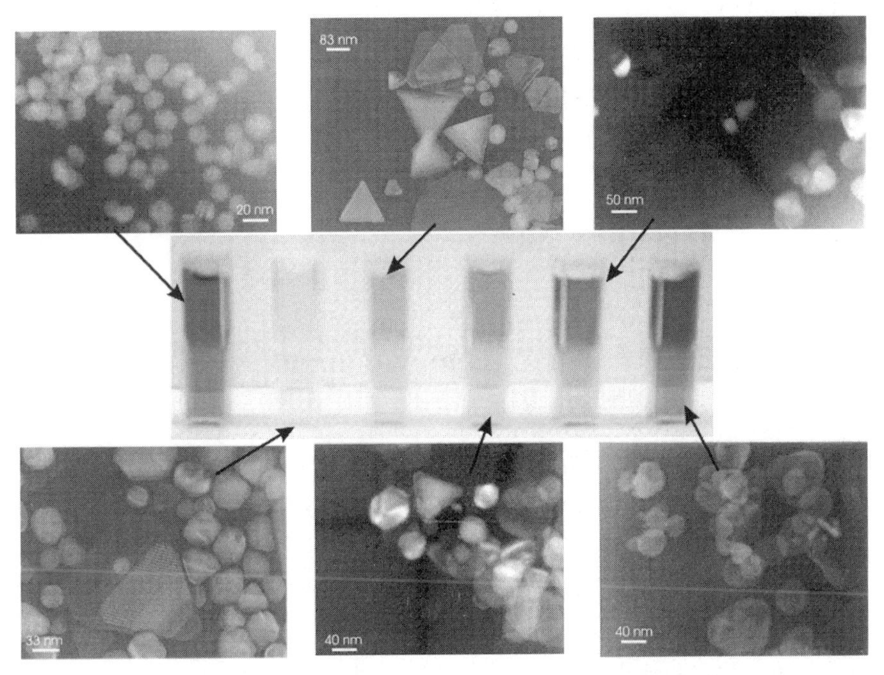

Figure 2.3
Nanocrystals in suspension. Each jar contains either silver or gold, and the color difference is caused by particle sizes and shapes, as shown in the structures above and below. *Courtesy of Richard Van Duyne Group, Northwestern University.*

close enough to combine, we can see the red color with the naked eye. That's how it works in the glass and glaze. If the dots are allowed to combine, however, they again look as golden as a banker's dream.

To understand why this happens, nanoscientists draw on information from many disciplines. Chemists are generally concerned with molecules, and important molecules have characteristic sizes that can be measured exactly on the nanoscale: they are larger than atoms and smaller than microstructures. Physicists care about the properties of matter, and since properties of matter at the nanoscale are rapidly changing and often size-controlled, nanoscale physics is a very important contributor. Engineers are concerned with the understanding and utilization of nanoscale materials. Materials scientists and electrical, chemical, and mechanical engineers all deal with the unique properties of nanostructures and with how those special properties can be utilized in the manufacturing of entirely new materials that could provide new capabilities in medicine, industry, recreation, and the environment.

The interdisciplinary nature of nanotechnology may explain why it took so long to develop. It is unusual for a field to require such diverse expertise. It also explains why most new nano research facilities are cooperative efforts among scientists and engineers from every part of the workforce.

SOME NANO CHALLENGES . . . .

Nanoscience and nanotechnology require us to imagine, make, measure, use, and design on the nanoscale. Because the nanoscale is so small, almost unimaginably small, it is clearly difficult to do the imagining, the making, the measuring, and the using. So why bother?

From the point of view of fundamental science, understanding the nanoscale is important if we want to understand how matter is constructed and how the properties of materials reflect their components, their atomic composition, their shapes, and their sizes. From the viewpoint of technology and applications, the unique properties of the nanoscale mean that nano design can produce striking results that can't be produced any other way.

Probably the most important technological advance in the last half of the 20th century was the advent of silicon electronics. The microchip—and its revolutionary applications in computing, communications, consumer electronics, and medicine—were all enabled by the development of silicon technology. In 1950, television was black and white, small and limited, fuzzy and unreliable. There were fewer than ten computers in the entire world, and there were no cellular phones, digital clocks, optical fibers, or Internet. All these advances came about directly because of microchips. The reason that computers constantly get both better and cheaper and that we can afford all the gadgets, toys, and instruments that surround us has been the increasing reliability and decreasing price of silicon electronics.

Gordon Moore, one of the founders of the Intel Corporation, came up with two empirical laws to describe the amazing advances in integrated circuit electronics. Moore's first law (usually referred to simply

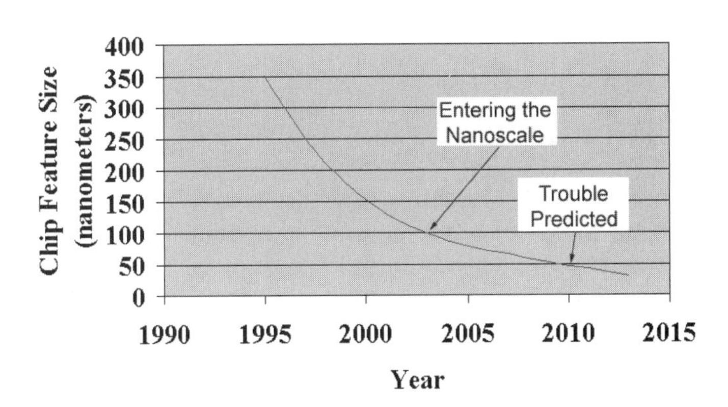

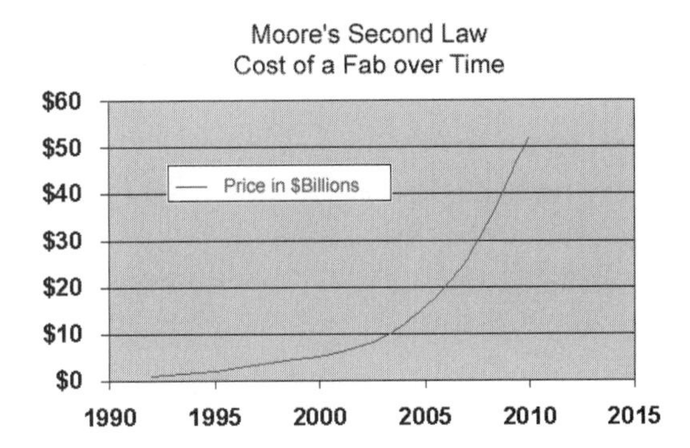

Figure 2.4
Moore's first and second laws.

as Moore's law) says that the amount of space required to install a transistor on a chip shrinks by roughly half every 18 months. This means that the spot that could hold one transistor 15 years ago can hold 1,000 transistors today. Figure 2.4 shows Moore's law in a graphical way. The line gives the size of a feature on a chip and shows how it has very rapidly gotten smaller with time.

Moore's first law is the good news. The bad news is Moore's second law, really a corollary to the first, which gloomily predicts that the cost of building a chip manufacturing plant (also called a fabrication line or just *fab*) doubles with every other chip generation, or roughly every 36 months.

Chip makers are concerned about what will happen as the fabs start churning out chips with nanoscale features. Not only will costs skyrocket beyond even the reach of current chip makers (multibillion-dollar fabs are already the norm), but since properties change with size at the nanoscale, there's no particular reason to believe that the chips will act as expected unless an entirely new design methodology is implemented. Within the next few years (according to most experts, by 2010), all the basic principles involved in making chips will need to be rethought as we shift from microchips to nanochips. For the first time since Moore stated his laws, chip design may need to undergo a revolution, not an evolution. These issues have caught the attention of big corporations and have them scrambling for their place in the nanochip future. To ignore them would be like making vacuum tubes or vinyl records today.

Aside from nanoscale electronics, one part of which, due to its focus on molecules, is often called *molecular electronics*, there are several other challenges that nanoscientists hope to face. To maintain the advances in society, economics, medicine, and the quality of life that have been brought to us by the electronics revolution, we need to take up the challenge of nanoscience and nanotechnology. Refining current technologies will continue to move us forward for some time, but there are barriers in the not too distant future, and nanotechnology may provide a way past them. Even for those who believe that the promise is overstated, the potential is too great to ignore.

3 Interlude One—The Fundamental Science Behind Nanotechnology

In this chapter...

Even though this book is meant to be for nonscientists, it's still helpful to review a few basic scientific principles before we dive into the dimensional home of atoms and molecules. These scientific themes come from physics, chemistry, biology, materials science, and engineering. We'll go over this material quickly, not making an attempt to deal with the sophistication and elegance that the science involves. This review is intended to be a user-friendly tour of the most significant scientific themes needed to understand the nanoscale. There are only two equations, we promise.

ELECTRONS

The chemist's notion of physical reality is based on the existence of two particles that are smaller than atoms. These particles are the *proton* and the *electron* (a *neutron* is effectively a combination of the two). While there are sub-subatomic particles (quarks and the like), protons and electrons in some sense represent the simplest particles necessary to describe matter.

The electron was discovered early in the 20th Century. Electrons are very light (2,000 times lighter than the smallest atom, hydrogen) and have a negative charge. Protons, which make up the rest of the mass of hydrogen, have a positive charge. When two electrons come near one another, they interact by the fundamental electrical force law. This force can be expressed by a simple equation that is sometimes called *Coulomb's law.*

For two charged particles separated by a distance r, the force acting between them is given as

$$F = Q_1 Q_2 / r^2$$

Here F is the force acting between the two particles separated by a distance r, and the charges on the particles are, respectively, Q_1 and Q_2. Notice that if both particles are electrons, then both Q_1 and Q_2 have the same sign (as well as the same value); therefore, F is a positive number. When a positive force acts on a particle, it pushes it away. Two electrons do not like coming near one another because "like charges repel" just as two north-polarized magnets do not like to approach each other. The opposite is also true. If you have two particles with opposite charges, the force between them will be negative. They will attract each other, so unlike charges attract. This follows directly from Coulomb's law.

It also follows from Coulomb's law that the force of interaction is small if the particles get very far apart (so that r becomes very big). Therefore, two electrons right near one another will push away from one another until they are separated by such a long distance that the force between them becomes irrelevant, and they relax into solipsistic bliss.

When electrons flow as an electrical current, it can be useful to describe what happens to the spaces they leave behind. These spaces are called "holes"; they aren't really particles, just places where electrons should be and are trying to get to. Holes are considered to have a positive charge; consequently, you can imagine an electric current as a group of electrons trying to get from a place where there is a surplus of electrons (negative charges) like the bottom of a AA battery to a place where there are holes (positive charges) like the top of a AA battery. To do this, electrons will flow through circuits and can be made to perform useful work.

In addition to forming currents, electrons are also responsible for the chemical properties of the atom they belong to, as we'll discuss next.

ATOMS AND IONS . . . .

The simplest picture of an atom consists of a dense heavy nucleus with a positive charge surrounded by a group of electrons that orbit the nucleus and that (like all electrons) have negative charges. Since the nucleus and the electrons have opposite charges, electrical forces hold the atom together in much the same way that gravity holds planets around the sun. The nucleus makes up the vast majority of the mass of the atom—it is around 1,999/2,000 of the mass in hydrogen, and an even greater percentage in other atoms.

There are 91 atoms in the natural world, and each of these 91 atoms has a different charge in its nucleus. The positive charge of the nucleus is equal to the number of protons it contains, so the lightest atom (hydrogen) has a nuclear charge of +1, the second lightest (helium) has a nuclear charge of +2, the third largest (lithium) has a nuclear charge of +3, and so forth. The heaviest naturally occurring atom is uranium, which has a nuclear charge of +92. (You might have guessed it was 91, but element number 43, technicium, does not occur naturally, so we skipped it.) You can see all of this on a periodic table.

In uncharged atoms, the number of electrons exactly balances the charge of the nucleus, so there is one electron for every proton. Hydrogen has one electron, helium has two, lithium has three, and uranium has 92. Since all the electrons are packed around the nucleus, generally the atoms with more electrons will be slightly larger than atoms with fewer electrons.

If the number of electrons doesn't match the charge of the nucleus (the number of protons), the atom has a net charge and is called an *ion* (also a favorite crossword puzzle word). If there are more electrons than protons then the net charge is negative and the ion is called a negative ion. On the other hand, if there are more protons than electrons, the situation is reversed, and you have a positive ion. Positive ions tend to be a touch smaller than neutral atoms with the same nucleus because there are fewer electrons, which are more closely held by the net positive charge. Negative ions tend to be a bit larger than their uncharged brethren because of their extra electrons. All atoms are roughly 0.1 nanometer in size. Helium is the smallest naturally occurring atom, with a diameter close to 0.1 nanometer, and uranium is the largest with a diameter of close to 0.22 nanometers. Thus, all atoms are roughly the same size (within a factor of 3), and all atoms are smaller than the nanoscale, but reside right at the edge.

These 91 atoms are the fundamental building blocks of all nature that we can see. Think of them as 91 kinds of brick of different colors and sizes from which it is possible to make very elegant walls, towers, buildings, and playgrounds. This is like the business of combining atoms to form molecules.

MOLECULES . . . .

When atoms are brought together in a fixed structure, they form a molecule. This construction resembles the way the parts are put together in children's building sets. Though there is a small set of parts, almost anything can be built within the confines of the builder's imagination and a few basic physical limits on how the parts fit together. Nature and the nanotechnologist have 91 different atoms to play with—each is roughly spherical but different in its size and its ability to interact with and bind to other atoms. Many, many different molecules exist—millions are known and hundreds of new ones are made or discovered each year. Figure 3.1 shows several molecules with from 2 to 21 atoms. All molecules with more than 30 or so atoms are more than a nanometer in size.

To form molecules, atoms bond together. There are a variety of types of chemical bonds, but they are all caused by interactions between the electrons of the atoms or ions involved. It isn't hard to see that a positive ion would be attracted to a negative ion, for example. We've already seen that attractive force at work in Coulomb's law. In fact, this is exactly the sort of attraction that forms the bonds in table salt (sodium chloride). The breaking and formation of bonds is a chemical reaction. Since electrons are responsible for bonds and since chemical reactions are just the making and breaking of bonds, it follows that electrons are responsible for the chemical properties of atoms and molecules. If you change the electrons, you change the

Figure 3.1
Models of some common small molecules. The white spheres represent hydrogen and the dark spheres represent carbon and oxygen. *From* Chemistry: The Central Science, *9/e, by Brown/LeMay/Bursten, © Pearson Education, Inc. Reprinted by permission of Pearson Education, Inc., Upper Saddle River, NJ.*

properties. Table salt is actually a good example of this. Both sodium and chlorine, the two atoms involved, are poisonous to humans if ingested individually. Combined, however, they are both safe and tasty.

Bonds are key to nanotechnology. They combine atoms and ions into molecules and can themselves act as mechanical devices like hinges, bearings, or structural members for machines that are nanoscale. For microscale and larger devices, bonds are just a means of creating materials and reactions. At the nanoscale, where molecules may themselves be devices, bonds may also be device components.

Smaller individual molecules are normally found only as vapors. When they mass together, molecules can interact with other atoms, ions, and molecules the same way that atoms can interact with each other, via electrical charges and Coulomb's law. Therefore, although an individual water molecule is a gas at room temperature, many water molecules clustered together can become a droplet of water, which is a liquid. When that liquid is cooled below 32°F (0°C), it becomes a solid. Liquid, solid, and gaseous water are all made of the same molecule, but the molecules are packed together in different ways.

Similar behaviors occur with many molecules. A carbon dioxide molecule normally forms a gas, but when many of these molecules cluster together, they form dry ice. Therefore, certain solid materials can be made simply of molecules. Usually these molecules are relatively small, consisting of fewer than a hundred atoms. Much larger molecules, called polymers, are materials by themselves and are key to nanoscience.

METALS

Most of the 91 naturally occurring atoms like to cluster with others of the same kind. This process can make huge molecule-like structures containing many billions of billions of atoms of the same sort. In most cases, these become hard, shiny, ductile structures called metals. In metals, some of the electrons can leave their individual atoms and flow through the bulk of the metal. These flowing electrons comprise electrical currents; therefore, metals conduct charge. Extension cords, power lines, and television antennas are all examples of devices where electrical charges move through metal structures.

This can be a little hard to imagine. Think of it as a bank where depositors are atoms, dollars are electrons, and the bank building itself is a macroscopic block of material or a huge molecule. You personally have a certain amount of money, which is probably pretty small in the grand scheme of the economy. However, once you deposit your money in a bank, it gets combined with all the money other people have deposited, and the money flows among the depositors and borrowers as needed. In case it gets lent to someone outside, it creates a business relationship with the borrower roughly analogous to a chemical bond. If you sever your relationship with the bank, you get to take your money with you, and, ignoring interest, you probably have the same amount you had when you arrived. The free flow of cash though this banking system is analogous to electrical current flowing through the bulk of our metal. The opposite case, where you keep your money under your pillow and there is no free flow or exchange, is analogous to electrical *insulators* or nonconductors. The analogy isn't perfect, but it may help.

Most metals are shiny because when light strikes a metal, the light is scattered by the moving electrons. Some materials are made of all the same atoms, but are not metallic. These materials tend to be made of lighter atoms. Some examples are graphite, coal, diamonds, yellow sulfur, and black or red phosphorus. They are sometimes called insulators because they do not have moving electrons to conduct charge. They are also generally not shiny because there are no free electrons to reflect the light that shines upon them. Even though we won't worry much about shininess, how free the flow of electrons in a material is matters quite a bit for nanotechnology.

OTHER MATERIALS

Nanoscience and technology focus on materials: physical and solid objects. Traditionally, materials science has been devoted to three large classes of materials—metals, polymers, and ceramics. We have just discussed metals, so let's look at the other two.

The most common polymers are plastics. They are sometimes called *macromolecules* to convey the sense that they are extremely large by molecular standards (though generally not big enough for a human to see individually, as the prefix "macro" would normally suggest). Most polymers are based on carbon because carbon has an almost unique ability to bond to itself. Polymers are single molecules formed of repeating patterns of atoms (called monomers) connected in a chain. In a sample such as a polystyrene drinking cup, there will be many different structures, and the chains will be of different lengths.

Polymers may be crosslinked, which means that the chains of monomers connect to other chains with bonds between the chains. Heavily *crosslinked polymers* not only tend to behave like the more conventional nonmetals but are also more likely to be harder because they have a rigid structure. The alternative is for the polymer chains to wrap and tangle like spaghetti or computer cables forming very pliable and rubbery materials. These are called *amorphous polymers*. Polyvinyl chloride (PVC), the material used to make pipes and a variety of other household goods, is an example of a heavily crosslinked polymer. Our polystyrene cup is mostly amorphous.

Simple polymers such as polyethylene or polystyrene are generally engineering plastics. Unlike the metals, carbon-based polymers are almost always insulating materials because the electrons remain localized near their parent atom's nucleus and cannot wander freely throughout the material. The fact that they are flexible insulators is also why plastics are used as jacketing for electrical wire. As might be expected, plastics are not shiny—think of a PVC shower curtain or polypropylene rope.

In addition to synthetic (man-made) polymers, there are many important polymers in the biological world. Examples include spider webs, the DNA molecules that store genetic information, proteins, and polysaccharides. These are discussed in the next section

Polymers generally do not conduct electricity, but it is possible to make special polymers that do. This fact is important because polymers are light, flexible, cheap, easy to make, and stable. For these reasons, using conducting polymers to replace metals in some applications, from low-tech applications such as static electricity prevention to nanoscience applications such as molecular wires, represents an important application of unusual polymers.

Figure 3.2
A molecular model of a segment of the polyethylene chain. This segment contains 28 carbon atoms (dark), but in commercial polyethylene there are more than a thousand carbon atoms per strand. *From* Chemistry: The Central Science, *9/e, by Brown/LeMay/Bursten, © Pearson Education, Inc. Reprinted by permission of Pearson Education, Inc., Upper Saddle River, NJ.*

The last area of traditional materials science is ceramics. *Ceramics* are often but not always oxides, which are structures where one of the atoms making up the extended structure is oxygen. Ceramics are made of several different kinds of atoms. Clay is mostly aluminum oxide, sand is mostly silicon dioxide, firebrick is magnesium silicon oxide, and calcium oxides are important in traditional tile applications. Like polymers and unlike metals, ceramics generally have localized electrons so they do not conduct electricity (though when super-cooled some can act as superconductors) and are generally not shiny. Ceramics are often very hard and sometimes brittle. They are only beginning to be used in nanoscience and nanotechnology, but they show promise for applications such as bone replacement.

So now we've discussed the three standard branches of materials science, but this discussion seems to leave out most of the materials with which we are familiar. A spade full of earth, a Western omelet, a loaf of bread, a meerschaum pipe, wood, fibers, and leaves are all *inhomogeneous structures* — they are made of many components, and the properties of the material reflect both the properties of those components and the unique properties that arise when the components are mixed. These inhomogeneous mixtures are very important for engineering applications, but for the most part they aren't very relevant at the nanoscale.

BIOSYSTEMS . . . .

Of the 91 naturally occurring elements, many are found in biology. As human beings, we require some highly unusual trace metals such as zinc, iron, vanadium, manganese, selenium, copper, and all the other goodies on the side of a vitamin jar for specific biological functions. Of the total weight of most plants and animals, however, well over 95 percent is made of four atoms: hydrogen, oxygen, nitrogen, and carbon. These are also the elements that dominate in most synthetic polymers. The reasons are quite straightforward. These atoms can form a wide variety of bond types; therefore, nature can use them to build some very complex nanostructures to accomplish the jobs of life, and scientists can use them to make new materials. For example, the molecules in our own bodies are responsible for respiration, digestion, temperature regulation, protection, and all the other jobs that the body requires. It clearly requires a wide assortment of fairly complex nanostructures to get the jobs done.

Generally, the molecules found in nature are complex and the source of much dismay to beginning organic chemistry students. For these molecules to perform useful functions, they must be easy to assemble and easy to recognize and bind to by other molecules. They must also be made by biological processes and have variable properties. To do this, these molecules are not usually simple repeating polymer structures such as polyethylene or polypropylene; instead, they are more complex irregular polymers.

There are four large classes of biological molecules. The first three are nucleic acids, proteins, and carbohydrates, which are all polymeric structures. The fourth catchall category is composed of particular small molecules that have special tasks to do.

Proteins make up much of the bulk of biology. Our nails and hair are mostly the protein keratin, oxygen is carried in our blood by the protein hemoglobin, and the protein nitrogenase is responsible for taking the nitrogen out of the air (on the nodules of legumes) and turning it into nitrates that permit plant growth. There are thousands of proteins, some of which are very well understood in terms of structure and function and some of which are still quite mysterious. Proteins are the machines of biology, the functional agents that make things happen.

Nucleic acids come in two categories called DNA and RNA. Both are needed to make proteins, but RNA has not yet been of major interest in nanostructures, so we'll only discuss DNA. A sketch of DNA is shown in Figure 3.3. It consists of a sugar outside containing negative charges due to the presence of phosphorous and oxygen atoms.

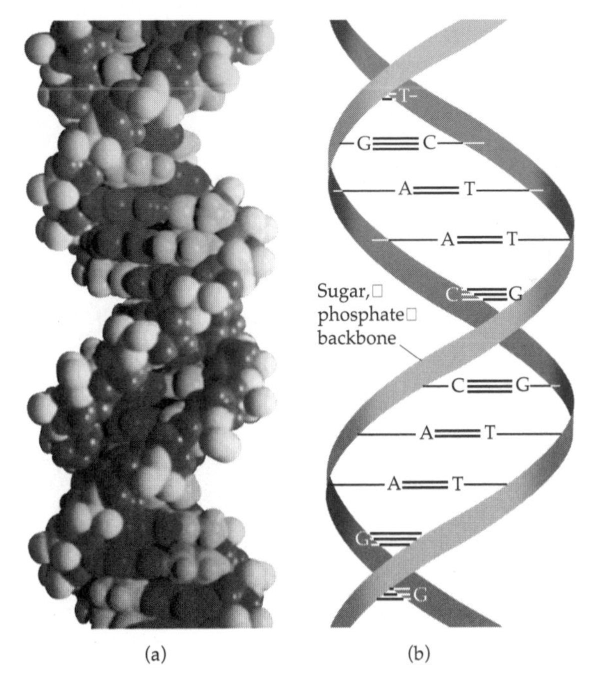

Figure 3.3
(a)Computer-generated model of the DNA double helix. b)Schematic showing the actual base pairs linked to each other. The light spheres represent hydrogen and the dark spheres represent carbon and oxygen. (*From* Chemistry: The Central Science, *9/e, by Brown/LeMay/Bursten, © Pearson Education, Inc. Reprinted by permission of Pearson Education, Inc., Upper Saddle River, NJ.*

Inside, there are stacked planar molecules that lie on top of one another like a pile of poker chips. Each of the poker chips consists of two separate planar molecules, held together weakly by bridges between oxygens or nitrogens and hydrogens. Because each poker chip is held at both its right and left ends, and because the structure is helical (a spiral), DNA has the structure of a double helix or double spiral staircase. It also looks (and to some extent acts) like a spring. When DNA is tightly wound, it is remarkably compact.

DNA is an almost unique molecule because each poker chip (called a base pair) can have one of four compositions (called AT, TA, CG, or GC). For each position on the strand, it is possible to control which base pair is present. That's because the two planar molecules that compose them can only be chosen from a set of four molecules called adenine, thymine, guanine, and cytosine, which are abbreviated A, T, G, and C. A and T will only bond to each other and not to G or C. Also, G and C will only bond to each other and not to A or T. Because of these limitations, the only possible base pairs are AT and GC and their opposites—TA and CG. These are placed on the double helix, in a particular order, and they code for all the functions of biology. The genetic code is simply an arrangement of base pairs in the DNA double helix, and it is a code that is read in a very sophisticated way by RNA and by proteins, which use the information to make protein-based biological structures that are the basis of life.

The third class of macromolecules found in biology is the *polysaccharides,* which are just sugars made of very long molecules. They are crucial to the functioning of the cell, and some of them are found in ligaments and in other biological structural materials. However, they are not yet of major use in synthetic nanotechnology.

The fourth class of biological molecules consists of very small molecules. These include water (crucial for the function of almost everything in biology), oxygen as a major energy source, carbon dioxide as the raw material for making plants, and nitric oxide. This last is a very small molecule consisting of a nitrogen and an oxygen linked together, and it plays many roles in biology from acting as "second messenger," a sort of relay messenger for communications within a cell, to causing erectile function.

There are other molecules that are less small but still crucial in biological applications. They include simple sugars and all drug molecules. Drugs generally work by binding either to a protein or to DNA and causing changes in those structures' functions. Sometimes the binding of these small molecules is very specific and very important.

MOLECULAR RECOGNITION . . . .

We've seen that molecules can have shapes and charges, and this means that parts of the molecule will be made of different atoms and will have different densities of electrons. Because Coulomb's law tells us that positive charges are attracted to negative charges, molecules can interact with one another by electrical (Coulombic) forces. For example, Figure 3.4 shows how charged atoms combine, and how two molecules can bind to each other based on the distribution of charge within the molecular structure.

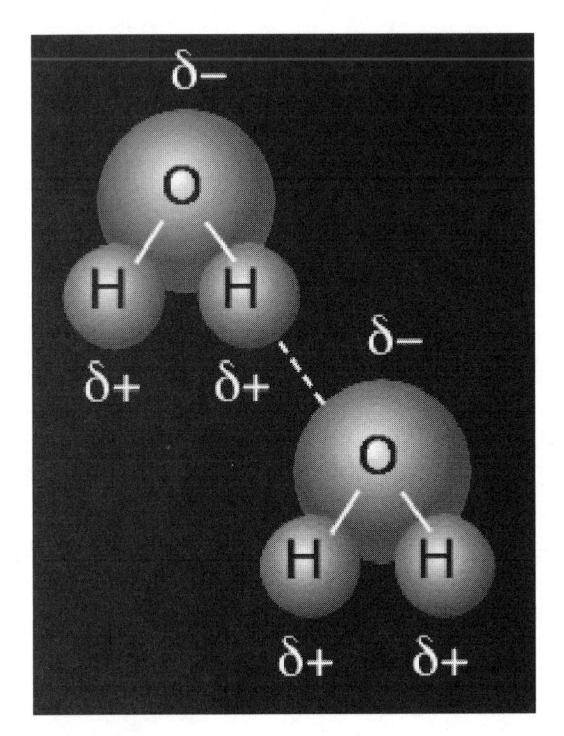

Figure 3.4
Molecular binding of two water molecules. The symbols δ+ and δ– denote positive and negative charges, respectively. *Courtesy of the Advanced Light Source, Lawrence Berkeley National Laboratory.*

The ability of one molecule to attract and bind to another is often referred to as *molecular recognition*. Molecular recognition can be very specific. It is the basic force in causing allergies, in which particular large molecules within the body recognize, bind to, and are affected by large foreign molecules, called allergens. These allergens include pollen, sugar, and some of the natural molecular components of chocolate, peanuts, and other things to which unfortunate people are sometimes allergic.

Molecular recognition can be used for other sensory experiences. Our sense of smell is based almost entirely on recognition of particular molecules by sensors in our nasal bulbs; consequently, molecular recognition underlies smelling a rose or newly cut grass. It can also identify smoke and keep you away from fire. Molecular recognition is also crucial in biology. Insects attract one another by manufacturing and emitting molecules called pheromones. If you are a frequent Internet user, you've probably gotten several email offers to buy human pheromones. Finally, molecular recognition can be used as a building strategy. Large biological molecules such as proteins can recognize one another and, in so doing, build the cells by which higher biological organisms are structured. Molecular recognition can cause a celery stalk to be stiff, water to quench our thirst, adhesives to stick, and oil to float on water.

Molecular recognition is one of the key features of nanotechnology. Because much of nanotechnology depends on building from the bottom up, making molecules that can organize themselves on their own or with a supporting surface like a metal or a plastic is a key strategy for manufacturing nanostructures. To give a macroscale analogy, if you want people to form a line, they must be able to see the line and where there is a place for them to stand. At the nanoscale, the job of "seeing" is done by molecular recognition.

ELECTRICAL CONDUCTION AND OHM'S LAW . . .

We usually use all our senses to become aware of objects. Light is seen with the eyes, pressure is felt in the ears and hands, and molecules are sensed in taste and smell. All these senses require an interaction between our bodies' sensory organs and external structures such as molecules or energy or physical objects.

The interactions that are important to taste, smell, and vision all require the flow of electrons within the body. Similarly, electrical charge moves through our nervous systems to inform the brain that a toe has been stubbed or a hand has gotten wet. All these signals, then, really rely on charge motion and, therefore, on Coulomb's law between like and unlike charges. Once again, all chemistry (and even biology) really boils downs to electrons. We know that metals contain free electrons that can move charge and reflect light. But even in non-metallic structures such as our nerves or our noses, electronic interactions and Coulombic forces are important. Moving electrons power our society, from light bulbs to batteries to computers.

Just as Coulomb's law is fundamental for describing the forces due to electrical charge, the current comprised of electrons moving through material also has a defining equation. This one is called Ohm's law.

The most common analogy for the flow of electrons is that of a river. Electron flow though a material is called current and is usually abbreviated as I and measured in electrons per second or a related unit. Resistance to the flow of current (analogous to rocks in the stream) is abbreviated as R. Voltage is the last of the key properties in Ohm's law and is the hardest to imagine. Voltage is the motive force that pushes the current along as the downward slope of a mountain watercourse pushes water. Voltage is abbreviated as V.

$$V = I R$$

Ohm's law, which simply states that voltage is equal to the current times the resistance, is obeyed in all the electrical and electronic circuits you deal with on a day-to-day basis. It isn't hard to see that this applies. If you have more motive force and the same amount of resistance, current should increase. If you keep motive force constant but increase resistance, current should drop. In almost all cases, this is true. Ohm's law works for hairdryers, computers, and utility power lines. All integrated circuits (chips) depend on Ohm's law.

But not everything obeys Ohm's law. Superconductors are materials in which there is effectively no resistance, and Ohm's law fails. There are other situations, including some special nanostructures such as carbon nanotubes, in which Ohm's law also fails. This leads to some interesting applications and challenges that we'll look at when we discuss molecular electronics.

QUANTUM MECHANICS AND QUANTUM IDEAS

Until the 20th Century, the physics of materials was dominated by Isaac Newton's ideas and formulas, which, with contributions over the next two centuries from many other notable scientists, formed the basis of classical mechanics. These laws describe fairly accurately all motion that you can see at a macroscale such as the movement of cars, the effect of gravity, and the trajectory of a punted football. But when physicists study very small structures at the nanoscale and below, some of the rules described in classical physics for materials fail to work as expected. Atoms don't turn out to behave exactly like tiny solar systems, and electrons show properties of both waves and particles. Because of these discoveries and many others, some of the ideas of classical mechanics were replaced or supplemented by a newer theory called *quantum mechanics*.

Quantum mechanics encompasses a host of interesting, elegant, and provocative ideas; however, for our current purposes, only a few significant notions are absolutely necessary. First, at these very small scales of length, energy and charge cannot be added continuously to matter but can only be added in small chunks. These chunks are called quanta (the plural of quantum) if they involve energy, and are units of electronic charge if they involve charge. Changing the charge on an ion, for example, can only be done by adding or subtracting electrons. Therefore, the charge of an ion is quantized (incremented) at the charge of one electron. There is no way to add half an electron.

Ordinary experience does not provide many examples of quantum behavior. Electrical current seems to be continuous, and the amount of energy that can be added to a soccer ball with a kick or a billiard ball with the strike of a cue seems to be continuously variable—the harder we push, the faster the ball moves. Despite this, there are some quantized things in common experience. One good example is money. You can't split a penny, but for amounts greater than one cent, you can always (theoretically) find cash to make exact change.

Many of the basic rules that define the behavior of nanostructures are the laws of quantum mechanics in disguise. Examples include issues such as how small a wire can be and still carry electrical charge, or how much energy we have to put into a molecule before it can change its charge state or act as a memory element.

OPTICS

Quantum mechanics can be significant for a number of issues involved in nanotechnology including understanding aspects of *optics*, how light interacts with matter. For example, the colors of individual dyes are fixed by quantum mechanics. The large molecule called phthalocyanine, which provides the blue color in jeans, can be changed to give greenish or purplish colors by modifying the chemical or geometric structure of the molecule. These modifications change the size of the light quanta that interact with the molecule and therefore change its perceived color. Similarly, different fluorescent lights give slightly more greenish or yellowish hues because the molecules or nanostructures that line the tube and emit light are changed. Even starlight has different colors, coming from stars of different temperatures and from different elements burning in the stellar atmosphere.

Light can also interact with matter in other ways. If you touch a black car on a sunny day, you will feel the heat energy that has been transferred to the metal by the light from the sun. Matter can also give off light energy as in fireworks and light bulbs. In all the cases that we are interested in, the total amount of energy involved in a process does not change (the technical term is that energy is conserved). But by manipulating this energy, we can cause very interesting things to happen.

As metallic objects become smaller, the quanta of energy (the sizes of the energy increments) that apply to them become larger. This relationship is similar to the behavior of drums: the tighter the drumhead, the higher the energy and pitch of the sound. It's also true of bells: generally, the smaller the bell, the higher the tone. This relationship between the size of a structure and the energy quanta that interact with it is very important in the control of light by molecules and by nanostructures and is a very significant theme in nanotechnology. It's also why our gold changed color in Chapter 2.

4 Interlude Two: Tools of the Nanosciences

[Nanofabrication] is building at the ultimate level of finesse.

Richard Smalley
Nobel Laureate and Professor, Rice University

In this chapter...

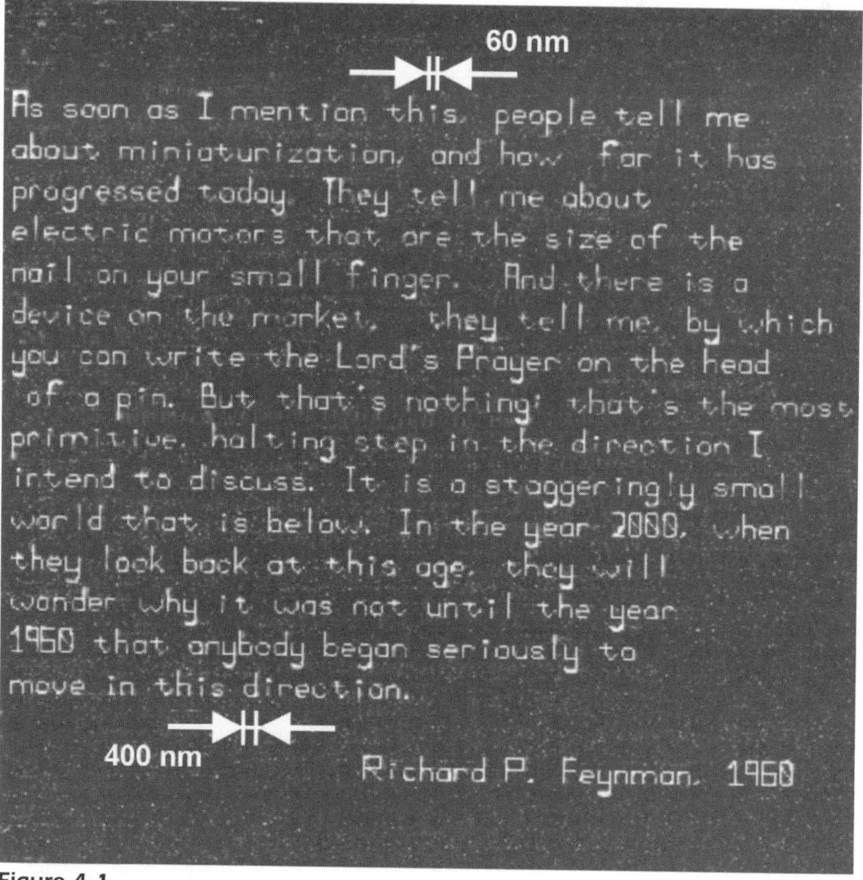

Figure 4.1
The founding speech of nanotechnology—written at the nanoscale.
Courtesy of the Mirkin Group, Northwestern University.

"In the year 2000, when they look back at this age, they will wonder why it was not until the year 1960 that anybody began seriously to move in this direction." (See Figure 4.1.) So said Nobel Prize–winning physicist Richard Feynman in a 1960 address commonly considered to have launched nanotechnology, but even he was a bit premature. While miniaturization continued at a breakneck pace, machines continued to shrink one step at a time in what we now call very prolonged top-down nanofabrication. No one immediately took up the challenge to start thinking from the bottom up, and it wasn't until the year 2000 (as Feynman predicted with uncanny accuracy) that devices started to break into the nanoscale and people started asking why we hadn't thought of this long before.

The reason is simple. We didn't have the tools. None of the manufacturing techniques that have allowed us to make smaller and smaller devices—microlathes, etchers, visible-light lithography equipment—are operable at the nanoscale. And not only couldn't we manipulate individual atoms and molecules, but we couldn't even see them until electron and atomic force microscopies were invented.

The reason why nanotechnology is coming to the surface now is that tools to see, measure, and manipulate matter at the nanoscale now exist. They are still crude, and the techniques with which we employ them are unrefined, but that is changing rapidly. It is now possible for a scientist in Washington, DC, using just an Internet connection to a remote-controlled laboratory in San Jose, California, to move a single atom across a platform in the lab. Technology continues to improve, and we have taken the, ahem, quantum leap into the nanoscale.

TOOLS FOR MEASURING NANOSTRUCTURES . . .

Scanning Probe Instruments

Some of the first tools to help launch the nanoscience revolution were the so-called scanning probe instruments. All types of *scanning probe* instruments are based on an idea first developed at the IBM Laboratory in Zurich in the 1980s. Essentially, the idea is a simple one: if you rub your finger along a surface, it is easy to distinguish velvet from steel or wood from tar. The different materials exert different forces on your finger as you drag it along the different surfaces. In these experiments, your finger acts like a force measurement structure. It is easier to slide it across a satin sheet than across warm tar because the warm tar exerts a stronger force dragging back the finger. This is the idea of the scanning force microscope, one of the common types of scanning probe.

In scanning probe measurements, the probe, also called a tip, slides along a surface in the same way your finger does. The probe is of nanoscale dimensions, often only a single atom in size where it scans the target. As the probe slides, it can measure several different properties, each of which corresponds to a different scanning probe measurement. For example, in *atomic force microscopy* (AFM), electronics are used to measure the force exerted on the probe tip as it moves along the surface. This is exactly the measurement made by your sliding finger, reduced to the nanoscale.

In *scanning tunneling microscopy* (STM), the amount of electrical current flowing between a scanning tip and a surface is measured. Depending on the way the measurement is done, STM can be used either to test the local geometry (how much the surface protrudes locally) or to measure the local electrical conducting characteristics. STM was actually the first of the scanning probe methods to be developed, and Gerd Binnig and Heinrich Rohrer shared the 1986 Nobel Prize for its development.

In *magnetic force microscopy* (MFM), the tip that scans across the surface is magnetic. It is used to sense the local magnetic structure on the surface. The MFM tip works in a similar way to the reading head on a hard disk drive or audio cassette player.

Computer enhancement is often used to get a human-usable picture from any scanning probe instrument, such as the nanoscale abacus that we saw in Chapter 1. It takes a great deal of enhancement just to make the raw results look as good as the ghostly x-ray pictures taken of your luggage at the airport. Scanning probe instruments can't image anything as large as luggage, however; they are more useful for measuring structures on length scales from the single atom level to the microscale. Nanotechnology will offer us other ways of catching baggage offenders.

Other types of scanning microscopies also exist. They are referred to as scanning probe microscopies because all are based on the general idea of the STM. In all of them, the important idea is that a nanoscale tip that slides or scans over the surface is used to investigate nanoscale structure by measuring forces, currents, magnetic drag, chemical identity, or other specific properties. Figure 4.2 shows an example of one of these tips.

Scanning probe microscopy made it possible to see things of atomic dimensions for the first time. It has been critical for measuring and understanding nanoscale structures.

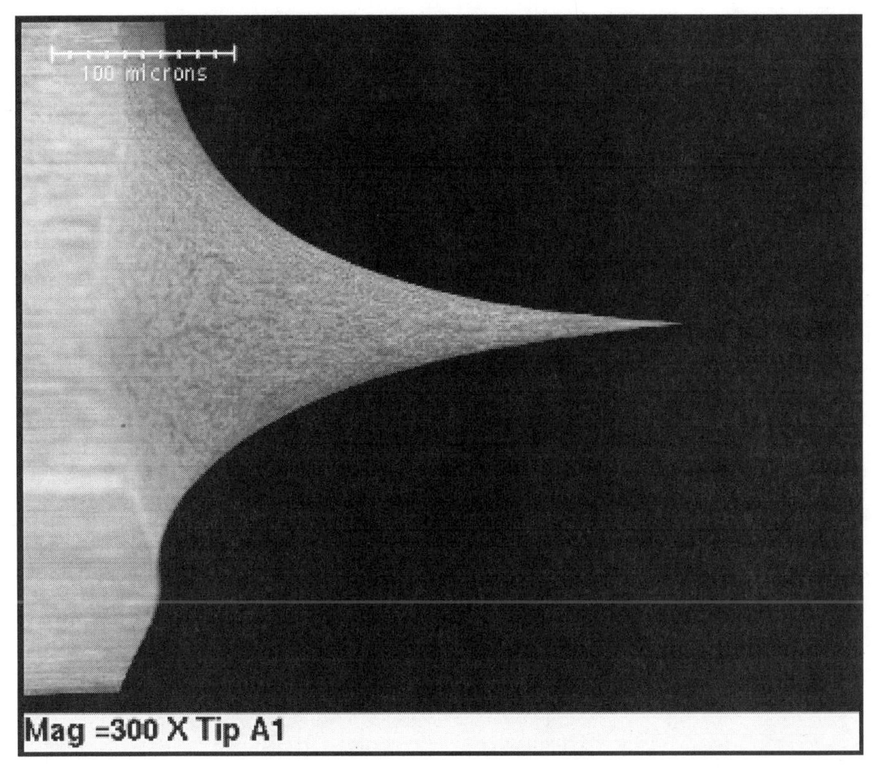

Figure 4.2
An STM tip made of tungsten. *Courtesy of the Hersam Group, Northwestern University.*

Spectroscopy

Spectroscopy refers to shining light of a specific color on a sample and observing the absorption, scattering, or other properties of the material under those conditions. Spectroscopy is a much older, more general technique than scanning probe microscopy and it offers many complementary insights.

Some types of spectroscopy are familiar from the everyday world. X-ray machines, for example, pass very high-energy radiation through an object to be examined and see how the radiation is scattered by the heavy nuclei of things like steel or bone. Collecting the x-ray light that passes through yields an image that many of us have seen in the doctor's office after a slip on the ice or in the bathtub.

Magnetic resonance imaging, or MRI, is another type of spectroscopy that may be familiar from its medical applications.

Many sorts of spectroscopy using different energies of light are used in the analysis of nanostructures. The usual difficulty is that all light has a characteristic wavelength and isn't of much use in studying structures smaller than its wavelength. Since visible light has a wavelength of between approximately 400 and 900 nanometers, it is clear that it isn't too much help in looking at an object only a few nanometers in size. Spectroscopy is of great importance for characterizing nanostructures en masse, but most types of spectroscopy do not tell us about structures on the scale of nanometers.

Electrochemistry

Electrochemistry deals with how chemical processes can be changed by the application of electric currents, and how electric currents can be generated from chemical reactions. The most common electrochemical devices are batteries that produce energy from chemical reactions. The opposite process is seen in electroplating, wherein metals are made to form on surfaces because positively charged metal ions absorb electrons from the current flowing through the surface to be plated and become neutral metals.

Electrochemistry is broadly used in the manufacturing of nanostructures, but it can also be used in their analysis. The nature of the surface atoms in an array can be measured directly using electrochemistry, and advanced electrochemical techniques (including some scanning probe electrochemical techniques) are often used both to construct and to investigate nanostructures.

Electron Microscopy

Even before the development of scanning probe techniques, methods that could see individual nanostructures were available. These methods are based on the use of electrons rather than light to examine the structure and behavior of the material. There are different types of *electron microscopy*, but they are all based on the same general idea. Electrons are accelerated and passed through the sample. As the electrons encounter nuclei and other electrons, they scatter. By collecting electrons that are not scattered, we can construct an image that describes where the particles were that scattered the electrons that didn't make it through. Figure 4.2 is a so-called transmission electron microscopy (TEM) image. Under favorable conditions, TEM images can have a resolution sufficient to see individual atoms, but samples must often be stained before they can be imaged. Additionally, TEM can only measure physical structure, not forces like those from magnetic or electric fields. Still, electron microscopy has many uses and is broadly used in nanostructure analysis and interpretation.

TOOLS TO MAKE NANOSTRUCTURES . . . .

The Return of Scanning Probe Instruments

Scanning probe instruments can be used not just to see structures but also to manipulate them. The dragging finger analogy is useful again here. Just as you can scratch, dimple, or score a soft surface as you drag your finger along it, you can also modify a surface with the tip of a scanning probe.

Scanning probes were used to manipulate the individual molecule beads on the molecular abacus in Figure 1.3. They have also been used to make wonderful nanoscale graffiti by arranging atoms or molecules on surfaces with particular structures. These structures have been used to demonstrate and test some fundamental scientific concepts ranging through structural chemistry, electrical interactions, and magnetic behaviors, among others. This assembling of materials on an atom-by-atom or molecule-by-molecule basis realizes a dream that chemists have had for many years.

Generally, small objects (which could be either individual atoms or individual molecules) can be moved on a surface either by pushing on them or by picking them up off the surface onto a scanning tip that moves around and puts them back down. For both cases, the scanning tip acts as a sort of earthmover at the nanoscale. In the pushing application, that earthmover is simply a bulldozer. In the pick-up mode, it acts more like a construction crane or backhoe.

Scanning probe surface assembly is inherently very elegant, but it suffers from two limitations: it is relatively expensive and relatively slow. It is great for research, but if nanotechnology is to become a real force, we must be able to make nanostructures very cheaply. (Recall our remarks concerning Moore's law, and the fact that silicon-based assembly methods have made transistors not only smaller but also cheaper and more reliable.) Although great advances have been made in building machines that use hundreds or even thousands of probe tips at the same time, making nanostructures using scanning probe tip methods is still very much like making automobiles by hand or blowing glass light bulbs individually. It can produce artistic and wonderful results, but it probably cannot be used to satisfy mass demand.

Nanoscale Lithography

The word "lithography" originally referred to making objects from stones. A *lithograph* is an image (usually on paper) that is produced by carving a pattern on the stone, inking the stone, and then pushing the inked stone onto the paper.

Many types of small-scale lithography operate in very much this way. Indeed, the common methods used to make current computer chips normally use optical or x-ray lithography, in which a master mask is made using chemical methods and light passes through that mask to produce the actual chip structures. It works just like a silk screen for a T-shirt.

Nanoscale lithography really can't use visible light because the wavelength of visible light is at least 400 nanometers, so structures smaller than that are difficult to make directly using it. This is one of the reasons that continuing Moore's law into the nanoscale will require entirely new preparation methods.

Despite this, there are several techniques for doing small-scale lithography. One of the most straightforward and elegant is *micro-imprint lithography*, largely developed by George Whitesides and his research group at Harvard. This method works in the same way as the rubber stamps that are still found in post offices. A pattern is inscribed onto a rubber surface (in this case actually a rubber-like silicon/oxygen polymer), and that rubber surface is then coated with molecular ink. The ink can then be stamped out onto a surface: this is paper in the post office, but it could be a metal, polymer, oxide, or any other surface in small-scale stamps. Small-scale stamping is more complex, but it is very inexpensive and can be used to make numerous copies. Originally, the stamps worked at the larger micron (1000-nanometer) scale, but recent improvements are bringing it to the nanoscale.

Dip Pen Nanolithography

One way to construct arbitrary structures on surfaces is to write them in exactly the same way that we write ink lines using a fountain pen. To make such lines at the nanoscale, it is necessary to have a nanopen. Fortunately, AFM tips are ideal nano-pens. *Dip pen nanolithography* (DPN) is named after the old-fashioned dip pen that was used in schoolrooms in the 19th century. The principle of DPN is shown in Figure 4.3, and the excerpt from Feynman's speech in Figure 4.1 is one DPN-assembled structure. In DPN, a reservoir of "ink" (atoms or molecules) is stored on the top of the scanning probe tip, which is manipulated across the surface, leaving lines and patterns behind.

DPN, developed by Chad Mirkin and his collaborators at Northwestern University, has several advantages, the two most

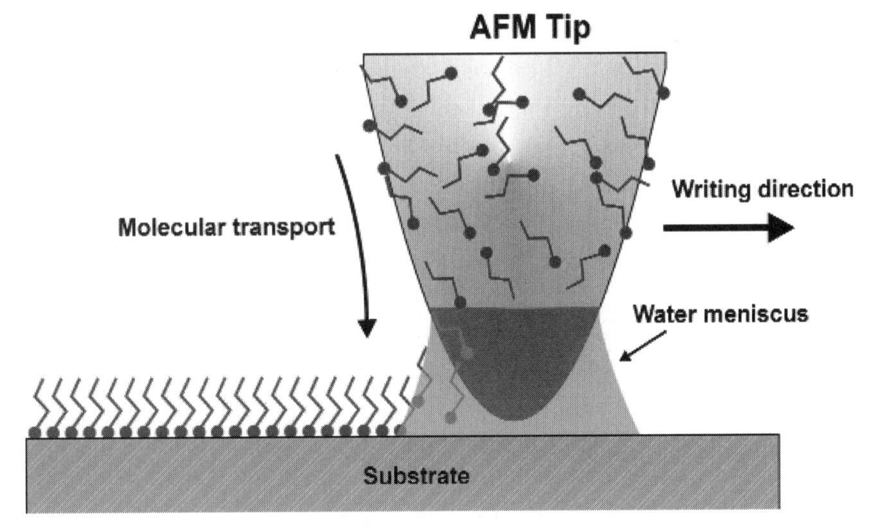

Figure 4.3
Schematic of the dip pen lithography process—the wiggly lines are molecular "ink." *Courtesy of the Mirkin Group, Northwestern University.*

important being that almost anything can be used as nanoink and that almost any surface can be written on. Also, you can use DPN to make almost any structure no matter how detailed or complex since AFM tips are relatively easy to manipulate. This fact makes DPN the technique of choice for creating new and complex structures in small volumes. The downside is that it is slow, unlike the nanostamp. It is like comparing hand illustration to early printing. Work is being done to improve this, notably by startup company NanoInk.

E-Beam Lithography

We mentioned that current light-based industrial lithography is limited to creating features no smaller than the wavelength used. Even though we can in principle get around this restriction by using light of small wavelengths, this solution can generate other problems. Smaller-wavelength light has higher energy, so it can have nasty side effects like blowing the feature you are trying to create right off the surface. (Imagine watering your garden plants with a fire hose.)

An alternate way of getting around the problem is to use electrons instead of light. This E-beam lithography can be used to make structures at the nanoscale. Figure 4.4 shows two electrodes that were made using E-beam lithography to align platinum nanowires. The structure lying across the nanoscale electrodes is a single molecule, a carbon nanotube.

E-beam lithography also has applications in current microelectronics manufacturing and is one approach that will be used to keep Moore's law on track until size-dependent properties truly assert themselves.

Nanosphere Liftoff Lithography

If marbles are placed together on a board as tightly as possible, they will form a tight group, with each marble surrounded by six others. If this array were spray painted from the top, and then the marbles were tipped off the board, the paint would appear as a set of painted dots, each shaped like a triangle but with concave sides (see Figure 4.5). Now if the marbles are nanoscale, so are the paint dots. In fact, Figure 4.5 shows dots of silver metal prepared by Rick Van

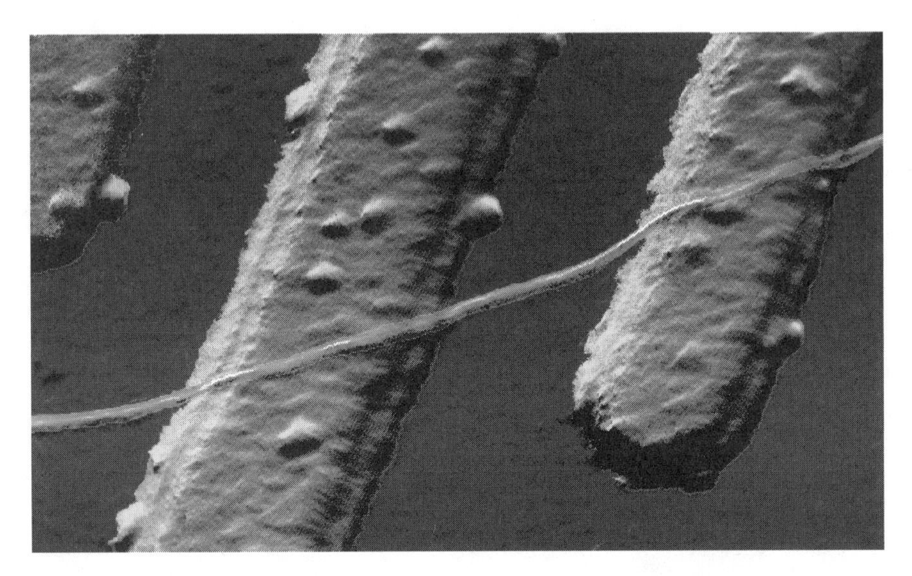

Figure 4.4
Two electrodes made using E-beam lithography. The light horizontal structure is a carbon nanotube. *Courtesy of the Dekker Group, Delft Institute of Technology.*

Duyne's group at Northwestern. The technique is called nanosphere liftoff lithography, even though no rockets are involved. It has several nice features: many sorts of boards (surfaces) and paints (metals, molecules) can be used, and several layers of paint (molecules) can be put down sequentially on the triangles. Importantly, this liftoff nanolithography, unlike DPN or scanning probe but like nanostamp, is parallel. Many nanospheres can be placed on the surface, so that regular arrays of many (thousands or more) dots can be prepared.

Molecular Synthesis

The production of molecules with particular molecular structures is one of the most active and wonderful parts of chemistry. Molecular synthesis involves making specific molecules for specific purposes, either with a purely scientific purpose or with very special application aims. There is extensive molecular synthetic work in drug companies, and many of the modern drugs including Penicillin, Lipitor, Taxol, and Viagra are the products of complex chemical synthesis.

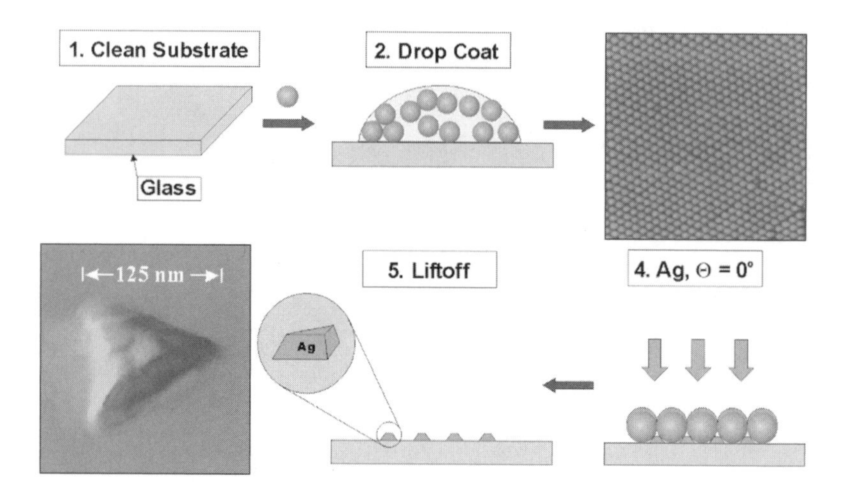

Figure 4.5
Schematic of the nanosphere liftoff lithography process. *Courtesy of the Van Duyne Group, Northwestern University.*

Making nanostructures with particular geometries at specific places on a surface means taking molecule making one step further. In addition to the chemical properties and composition of a molecule, *nanoscale synthesis* must also be concerned with the physical layout and construction of nanostructures. For example, some of the drug delivery techniques we'll look at later involve taking active elements from drugs and pushing them into nanoscale shells to allow them to pass into areas of the body where they could not penetrate before. To do this, the drug must be injected into the molecular shell like jelly into a donut. There is only a physical interaction here; there are no chemical bonds between the two.

Any technique involving manipulating atoms one by one is clearly too slow and cumbersome, especially if we wish to make bulk materials or even enough of our encapsulated drug to treat a person.

Self-Assembly

The problem with most of the techniques for assembling nanostructures that we've seen so far is that they are too much like work. In every case, we try to impose our wills on these very small objects and manipulate and tweak them to be just how we want them. Wouldn't it be glorious if we could just mix chemicals together and get nanostructures by letting the molecules sort themselves out?

One approach to nanofabrication attempts to do exactly this. It is called *self-assembly*. The idea behind self-assembly is that molecules will always seek the lowest energy level available to them. If bonding to an adjacent molecule accomplishes this, they will bond. If reorienting their physical positions does the trick, then they will reorient. At its simplest, this is the same underlying force that causes a rock to roll down a hill. No matter how you lift, throw, twist, crush, or manipulate the rock, it will always try to get down the hill. You can block its progress, but that requires active intervention. In this case, the rock is trying to minimize its gravitational energy. In the case of a molecule, it is trying to minimize other kinds of energies. Thanks to Coulomb's law, these are most frequently forces from charge interactions.

One way to imagine self-assembly is to imagine a compass. If you shake it, you can cause the needle to fluctuate and point in almost any direction for an instant, but once you stop shaking it, the needle will ultimately reorient itself and point from south to north. There is a small magnet in the needle, and this south-to-north orientation minimizes its energy with respect to earth's magnetic field. You don't need to do any work on the needle to get it to do this. It does it naturally. Self-assembly techniques are based on the idea of making components that, like our compass needle, naturally organize themselves the way we want them to.

The forces involved in self-assembly are generally weaker than the bonding forces that hold molecules together. They correspond to weaker aspects of Coulombic interactions and are found in many places throughout nature. For example, weak interactions called hydrogen bonds hold the hydrogen atom in one molecule of liquid water together with the oxygen atom of the next and prevent the molecules from becoming water vapor at room temperature.

Hydrogen bonding also helps to hold proteins into particular three-dimensional structures that are necessary for their biological function.

Other weak interactions also exist, including the hydrophobic interactions that allow oil to float on water and multipolar interactions. Multipolar interactions occur between structures, each of which has no total charge (so it is not like an electron interacting with another electron, which is a strong Coulombic interaction). Rather, here there are different distributions of charge on two molecules interacting with one another. Such multipole interactions are generally weak, but they are strong enough to provide very complex structures.

In self-assembly, the nano builder introduces particular atoms or molecules onto a surface or onto a preconstructed nanostructure. The molecules then align themselves into particular positions, sometimes forming weak bonds and sometimes forming strong covalent ones, in order to minimize the total energy. One of the huge advantages of such assembly is that large structures can be prepared in this way, so it is not necessary to tailor individually the specific nanostructures (as was true in AFM, STM, and DPN construction of nanoscale objects). Self-assembly is almost certainly going to be the preferred method for making large nanostructure arrays, such as the computer memories and computer logic that must be prepared if Moore's law is going to continue to hold true beyond the next decade.

Self-assembly is not limited to electronics applications. Self-assembled structures can be used for something as mundane as protecting a surface against corrosion or making a surface slippery, sticky, wet, or dry. Figure 4.6 shows some great examples of self-assembly from the laboratory of Sam Stupp at Northwestern. In this case, two levels of self-assembly are used. First, the self-assembly of long complex molecules called rodcoils produces the mushroom-like nanostructure. Then the nanostructures themselves self-assemble to produce a surface coating that makes the glass slide either *hydrophilic* (water loving—easily wetted) or *hydrophobic* (water hating—so that the water beads up). This also shows that very complex structures can be formed using self-assembly by breaking tasks down into steps.

Self-assembly is probably the most important of the nanoscale fabrication techniques because of its generality, its ability to produce structures at different length scales, and its low cost.

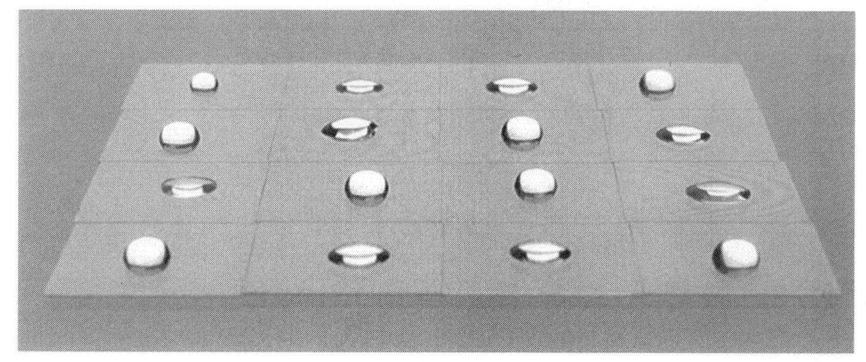

Figure 4.6
Molecular model (top) of a self-assembled "mushroom" (more correctly a rodcoil polymer). The photograph (bottom) shows control of surface wetting by a layer of these mushrooms. *Courtesy of the Stupp Group, Northwestern University.*

Nanoscale Crystal Growth

Crystal growth is another sort of self-assembly. Crystals like salt that are made of ions are called, unsurprisingly, ionic crystals. Those made of atoms are called atomic crystals, and those made of molecules are called molecular crystals. So salt (sodium chloride) is an ionic crystal, and sugar (sucrose, $C_{12}H_{22}O_{11}$) is a molecular crystal.

Crystal growth is partly art, partly science. Crystals can be grown from solution using seed crystals, which involves putting a small crystal into the presence of more of its component materials (usually in solution) and allowing those components to mimic the pattern of the small crystal, or seed. *Silicon boules*, the blocks used for making microchips, are made or "drawn" in this way.

By making clever choices of seed crystals and growing conditions, it is possible to cause the crystals to assume unusual shapes. Charles Lieber and his group at Harvard University have used nanoscale crystals to seed long, wire-like single crystals of carbon nanotubes as well as compounds such as indium phosphide or gallium arsenide, and of atomic crystals such as silicon. These nanowires (one is shown in Figure 4.7) have remarkable conductivity properties, as well as many uses both in optics and in electronics.

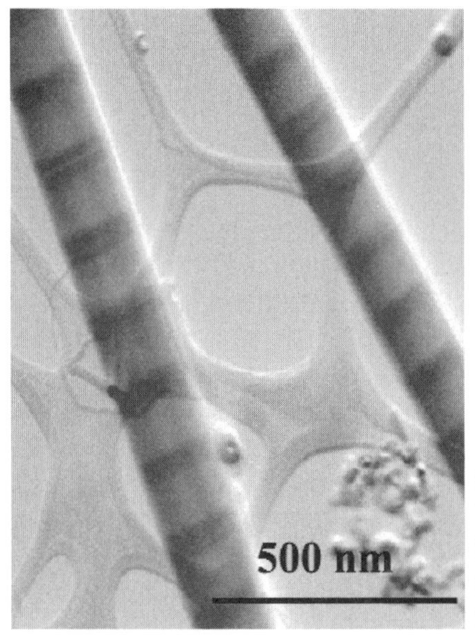

Figure 4.7
Two parallel nanowires. The light color is silicon, and the darker color is silicon/germanium. *Courtesy of Yang Group, University of California at Berkeley.*

Polymerization

As we discussed in Chapter 3, polymers are very large molecules. They can be upward of millions of atoms in size, made by repetitive formation of the bond from one small molecular unit (monomer) to the next. Polymerization is a very commonly used scheme for making nanoscale materials and even much larger ones—epoxy adhesives work by making extended polymers upon mixing the two components of the epoxy.

Ordinarily, industrial polymers like polystyrene or polyethylene or polyvinylchloride (PVC) are made by building extremely long molecules, with numerous steps that occur sequentially. *Controlled polymerization*, in which one monomer at a time is added to the next, is very important for specific elegant structures. Robert Letsinger and his students at Northwestern University have developed a series of methods for preparing specific short DNA fragments. These are called *oligonucleotides* from the Greek work "oligo," which means a few. (A monomer is one unit, an oligomer is several units, and a polymer is many units.) The so-called gene machines use elegant reaction chemistry to construct specific DNA sequences.

Building specific DNA sequences is crucial for many reasons. In modern biotechnology, these specific sequences are used to build new biological structures (drugs, materials, proteins), based on the ability of bacteria to reproduce themselves. A synthetic DNA template is introduced into the bacterial DNA, and the bacteria then produce many copies of that particular target protein. The modification of the bacteria's DNA is done using a series of chemical reactions, and the gene machines are used to prepare the specific short oligonucleotides to modify bacterial DNA, capturing that process to produce the protein of choice. This allows you to effectively make protein factories for nearly any protein you choose. One good example of how this could be used is to make the protein insulin for the treatment of diabetes.

The combination of specific short DNA sequences and self-assembly is used extensively to make materials in which a single DNA strand binds to another single DNA strand. This process, called *hybridization*, is shown in Figure 4.8. Recall that the DNA base A always pairs with T, and the DNA base G always pairs with C. In Figure 4.8, the perfect matching on the left gives a tighter, stronger fit than the imperfect mismatched set. This kind of self-assembly is pre-

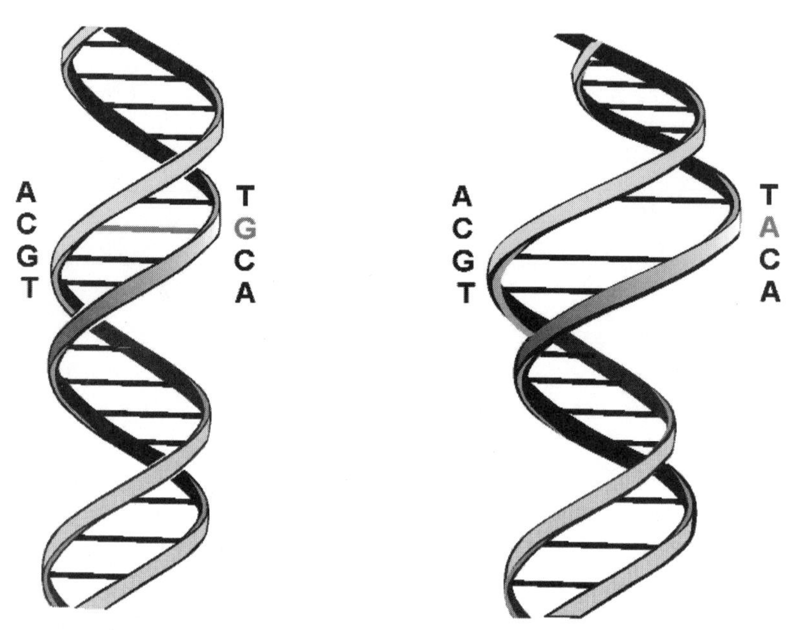

Figure 4.8
Schematic of the DNA hybridization process. The "matched" side shows how a DNA strand correctly binds to its complement and the "mismatched" side shows how errors can prevent binding. *Courtesy of the Mirkin Group, Northwestern University.*

sent in nature—it's how DNA replicates so that cells can multiply. Many synthetic applications of this complementary molecular recognition are used in nanoscience.

Nanobricks and Building Blocks

Nanostructures must be assembled from components. The fundamental building blocks are atoms of the 91 naturally occurring elements. Usually, though, it is inefficient to start with individual atoms. We saw both the strength and the slowness of this approach when we discussed building atomic scale nanostructures using scanning probe microscopy, especially if we are trying to make a macroscopic amount of a material rather than build a single nanoscale machine. Richard Smalley, who won the Nobel Prize for his nanoscience work in 1996, estimated that it could take nanomachines as much as 19 million years to build a few ounces of material building atom by atom because the number of atoms in such a sample is about 6-with-23-zeros-after-it. If one atom were the size of a teaspoon full of water, this many atoms would be about the size of the Pacific Ocean.

Building an ocean one teaspoon at time would be a very slow process and so is building bulk materials atom by atom. Assembling at the rate of a million atoms per second would still take 6-with-17-zeros-after-it seconds to construct a handful of useful material. (For comparison, the national debt of the United States is currently approximately 6-with-12-zeros-after-it dollars.) This is a bit of a damper for those who imagine nanoscale robots (sometimes called "assemblers") running around making everything from cars to clocks, but there are already promising alternatives for making bulk materials based on nanostructures.

Usually, nanostructures are built starting with larger building blocks or molecules as components. You might think of these as nano Legos. Sometimes these are traditional small molecules. The weak interaction of the sulfur group with the gold surface is often used to construct beautifully adherent, adhesive, and regular films of sulfur-ended long molecules on the gold surface. These molecules are called alkane thiols. "Alkane" means a long chain of carbon–carbon bonds of exactly the same sort seen in polyethylene. The "thiol" refers to the sulfur on the end that links (self-assembles) onto the gold surface to form the monolayer. The monolayer can be nanometers thick and very large in the other two dimensions. It is built not from the individual atoms but from the alkane thiol molecules on the gold surface. The National Nanotechnology Initiative quote in Figure 1.1 was written on gold with alkane thiol ink using dip pen nanolithography.

In addition to individual molecules of the sort that would be found in traditional chemistry laboratories, some very new semimolecular building blocks are used to assemble nanostructures. Two of these nanostructures are so-called carbon *nanotubes* (first prepared by Sumio Iijima in Tokyo) and *nanorods* that can be made out of silicon, other semiconductors, metals, or even insulators. These nanorods are made using clever solution chemistry methods, but they can then self-assemble into larger nanoscale structures.

Common Threads of Nanotechnology: Nanotubes and Nanowires

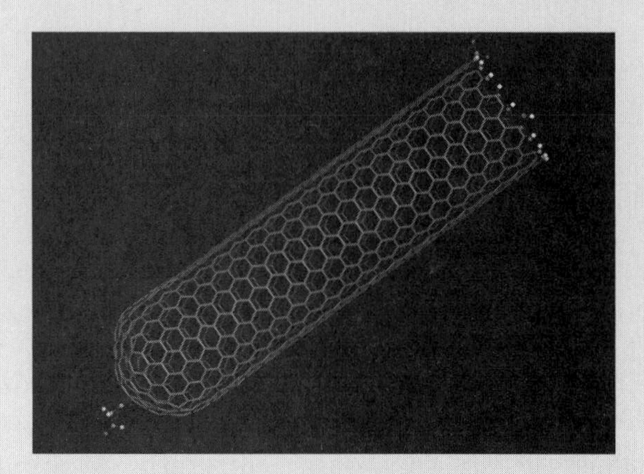

Figure 4.9
A singled-walled carbon nanotube. *Courtesy of the Smalley Group, Rice University.*

Everyone is familiar with graphite as the black stuff in pencils. Some people also use it as a lubricant for machinery because, at the molecular level, it is formed out of sheets of carbon that slide across each other with very little friction. These sheets of carbon are comprised of carbon atoms linked together hexagonally like chicken wire. Nanoscientists are very interested in them because, when rolled into tubes, they have some amazing properties. These cylinders of graphite are called carbon nanotubes. When the roll is only one sheet of carbon atoms thick, they are called single-walled carbon nanotubes. Nanotubes are some of the first true nanomaterials engineered at the molecular level, and they exhibit physical and electrical properties that are truly breathtaking.

Estimates vary as to precisely how strong a single carbon nanotube can be, but laboratories have already shown their tensile strength to be in the excess of 60 times stronger than high-grade steel. By some estimates, a nanotube fiber narrower than a human hair might be able to suspend a semitrailer, though no one has yet been able to make a tube big enough to find out for sure. Others guess that even a single tube could stretch from earth to the stratosphere and be able to support its own weight. Many nanoscientists assert that nanotubes are not only the strongest materials ever made, they are among the strongest materials it will ever be possible to make. Absolute statements of this type are now possible since nanotechnology allows for the possibility to engineer at the ultimate level of detail, building designer materials atom by atom rather than building large-grained composites like concrete and plywood.

Not only are nanotubes very strong, but they are also light and flexible. Other forms of carbon such as carbon fiber are already used in high-end sports equipment and airplane design because they have strengths comparable to steel or aluminum at a fraction of the weight. Nanotube materials could take this to the next level, but nanotube fabrication is still in its infancy and even the most sophisticated factories still produce only grams of nanotubes per week. Because of this, nanotubes are hard to get and they are also very expensive. Until manufacturing technology improves, the application of the amazing physical properties of these materials will be extremely limited, though the first few nanotube-enhanced products are beginning to hit the shelves in the form of tennis rackets and golf clubs. In these products, small amounts of nanotubes are introduced into a more traditional material to form a composite. These nanotube-composite products don't take full advantage of the potential of nanotubes, but they do hint provocatively at how much potential nanotubes have.

The physical properties of nanotubes are striking enough, but their electrical properties may be even more exciting. Looking at the shape of a nanotube, nanoscientists predicted that electrons might be able to shoot up and down the tube using it as a wire. When this was tested, some scientists found that they acted almost like superconductors, transmitting electricity without resistance. Others found that they acted like semiconductors. Current theory holds that they can act as either superconductors or semiconductors, depending on the exact proportions of the tube and which materials other than carbon are introduced into the tube matrix (a process called *doping*).

Not all nanotubes are made out of carbon. Silicon nanotubes are also common, though noncarbon tubes are usually called *nanowires*. The versatile electrical properties of these nanotubes and nanowires are now being explored for purposes of making nanoscale electronic devices. Nanotubes are approximately 1 percent of the size of connecting elements on current state-of-the-art microchips, and the idea of superconducting interconnects offers the tantalizing possibility of overcoming one of the greatest bugaboos of current chip design—the waste heat that is created as electrons flow through metal wires.

Nanotube and nanowire research and manufacturing is a hot topic both for scientists and industry. Several startup companies have been created to make them, and these companies are finding a ready market for their products. Engineers in Phaedon Avouris's group at IBM have already used nanotubes to craft usable transistors with properties exceeding those of their pure silicon cousins, and some nanotube-based logic gates have been produced, opening one avenue to nanoscale computing. Nanotubes and nanowires are not known to occur naturally and are among the early returns on investment in nanoscience and technology.

Tools to Imagine Nanoscale Behaviors

Synthetic chemistry is a science that uses well-understood principles to construct large molecular structures. Understanding how to construct nanoscale materials and what the properties of those nanoscale materials are is central to nanoscience. As we have stressed in previous chapters, nanoscale materials have properties quite different from materials that exist at the atomic scale (simple gasses) or from extended structures such as metals, polymers, or ceramics. Conceptual, modeling, and theoretical ideas about nano behavior are crucial in order to develop techniques for mass production at the nanoscale.

The fundamental ideas involved are the ones we talked about in Chapter 3—electrical interactions and Coulomb's law, the behaviors and design rules of quantum mechanics, interactions with light, and interactions among components, largely based on manifestations of the Coulombic force. Therefore, the traditional methods of theoretical physics, theoretical chemistry, electrical engineering, and materials science dominate both the concepts used to understand nanostructures and the calculations that are necessary to predict the behavior. They suggest the design of nanoscale structures and devices.

The use of quantum mechanical theory to predict actual molecular structures has been one of the triumphs of chemistry in the 20th Century. Extension of these models to deal with larger nanostructures is moving very rapidly. These calculations are done using large computers and can be mixed with more traditional predictions based on classical mechanics. The computer predictions of structures are one way by which nanoscale design occurs, although they are most useful when mixed with intuition, experience, and inspiration.

NanoCAD

Complementary metal oxide semiconductor (CMOS) silicon chip making technology remains one of the key methods for making microchips such as computer microprocessors. In design of CMOS structures, the only fundamental scientific requirement is the behavior of currents that are described by Ohm's law. Still, assembling a very complex CMOS structure such as that shown in Figure 4.10 is extremely complicated both physically and financially (making the

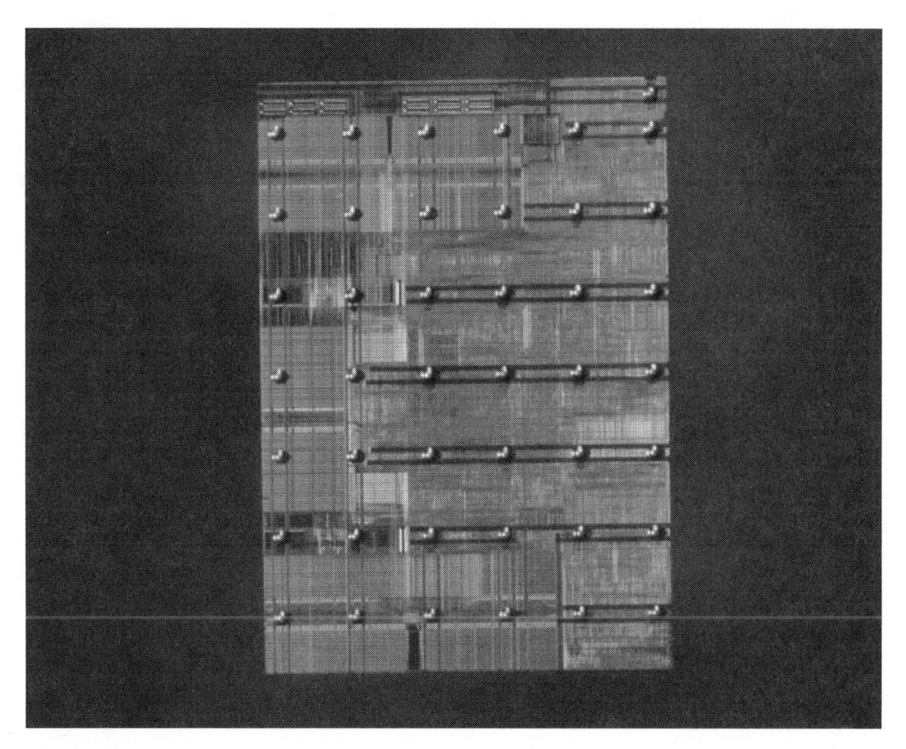

Figure 4.10
A current CMOS chip surface. *Courtesy of Tom Way/IBM Corporation.*

tools and the masks, doing the lithography, and assembling millions of such chips on tiny amounts of chip area are very hard to do for a low price). To make the modern chips of the types used in consumer electronics, it is necessary to assemble millions of transistors on silicon chips and to route charge through them in such a way as to perform the algorithms or calculations that the chip is designed to carry out. In the early days of chip design, engineers would literally sketch the circuits they wanted on the boards, and companies maintained art departments of highly coordinated individuals to draw these features precisely (by hand, with a pen under a magnifying glass) as templates for making masks. In those days, though, chips contained only a few dozen components. Designing chips with millions of components is simply too complicated to be done by individuals.

Due to these issues, computers design most functional chips. Computer programs, generically called Spice or Computer-Aided

Design (CAD), are specifically programmed to design chips for particular functions. Engineers design the chips at a high level using specialized hardware description computer languages or graphical environments that use predesigned components.

The CAD programs are used to design structures on scales far larger than that of a nanometer. Typical CMOS structures are still hundreds or thousands of nanometers in size, as they must be. When structures fall into the nanoscale regime, Ohm's law doesn't necessarily work. The phenomena of charge motion are then described by quantum mechanics, and the understanding of how currents will flow becomes much more complicated. Standard transistors behave quite differently than those that one would anticipate at the nanoscale, and there are not yet any nanoscale Spice or CAD programs. Several laboratories are addressing this challenge, but until we understand the behavior of individual nanostructures well enough, building a Spice-like computer code for assembling nanodevices into logic structures and architectures will remain a tantalizing challenge.

The Swarm Mentality: Amorphous Computing

Since the 1970s, microchips have gotten much faster and much more complex. Modern processors, for example, can add numbers that are 64 bits (more than 19 decimal digits) in a single step, whereas early computers could only work with eight-bit numbers (fewer than three decimal digits). To support high-level programming languages, dozens of microcode instructions must be programmed into the hardware of the chip. Intel's code for its Pentium chips, for example, contains around 100 commonly used instructions. This means that modern processors must have millions of components and complex instruction queuing (often called *pipelining*). Their architecture reflects 20 years of refinements, hacks, and complexities.

Although molecular electronics is likely to be commercially viable for many applications, many scientists and engineers think it is unlikely that devices as complex as a 10-million-transistor Pentium-style processor are realistic in the next few years, especially given the slow and crude assembly techniques currently available. What is more likely is that first-generation molecular "nanoprocessors" will be very simple, capable of performing just a few basic behaviors such as "generally follow your neighbor processors" and "switch quickly to the next position." Again, because of the tolerances and techniques involved, each processor won't be able to perform its task with digital exactness and certainty, but if a processing cluster contains millions or even billions of such molecules, many types of errors will be eradicated in the averages.

This is, to a great extent, how communities work. Ants don't have a single central processor telling them how to build an anthill, nor do bees for a hive or people for a city. Anthills, hives, and cities don't fail because a single creature couldn't build them, and they don't fail to perform as expected even in the face of plagues, earthquakes, or errant picnickers. You have to destroy the overwhelming majority to destroy the whole. If computers could be built in the same way, then the question of scaling computing power becomes as easy as adding processors. Overall reliability and system integrity would be increased, but it would require fundamentally rethinking the way computers are made and programmed today. The computer scientists and engineers that are pursuing this study call the discipline *amorphous computing* or *swarm computing* after its relationship to swarming insects. Though not strictly nanoscience, this sort of research may be a key to molecular or quantum computing.

5 Points and Places of Interest: The Grand Tour

In this chapter...

Nanoscience and nanotechnology have become a dominant theme in many research institutions worldwide. Efforts in the United States and several other countries have led to the development of nanoscience centers, which are generally situated at major universities. In this chapter, we will take a short tour around some of the major development areas in nanoscience and nanotechnology. The grand tour could have been completed in any of many centers worldwide, but we will focus on research being done in the Center for Nanofabrication and Molecular Self-Assembly at Northwestern, which is the first federally funded nanotechnology and nanoscience center in the United States. Because nanotechnology and nanoscience are done in government, industrial, and academic laboratories throughout the world, this tour will also feature stops in major industrial labs and one exciting laboratory in Europe.

SMART MATERIALS . . . .

Michael Wasielewski is Chair of the Chemistry Department at Northwestern. He is a Chicago native, with a deep affinity for the Cubs, and spent several years working at Argonne National Laboratory before moving to Northwestern. One of Wasielewski's major research interests is a complex set of polymeric materials generally referred to as *photorefractive polymers*. These highly unusual structures contain mobile electronic charges, almost like metals. The mobile charges can be moved to new positions either by shining light on the polymers or by putting them in an electric field. The position of these charged particles is then a sort of code, a code that can be read by shining different colors of light on the coded polymer, making it work much like a nanoscopic version of a supermarket barcode reader. Photorefractive polymers are of major interest as information storage devices whose storage density can far exceed even the best available magnetic storage structures.

Photorefractive polymers are a particularly complicated and a wonderful form of nanoscale *smart materials*. In nanoscience, the term "smart material" refers to any material engineered at the nanoscale to perform a specific task. Sometimes smart materials are also dynamic, which means that the material can change its most basic properties or structure based on an outside cue. A simple example of a dynamic smart material is self-tinting automotive glass that is clear most of the time but darkens under intense light to prevent blinding a driver. In the case of photorefractive polymers, the ability to move charges using light or an electric field is engineered into the material at its most basic level. No materials could be made to do this without nanotechnology and manipulation at the nanoscale.

Chapter 6 deals more generally with the topic of smart materials, how they are designed and manufactured, and some of the applications envisioned for them.

SENSORS

Joe Hupp teaches chemistry at Northwestern University and, in his short scientific career, has worked in many different areas of chemistry and materials. He is athletic, quiet, intense, brilliant, and youthful. One of Joe's major areas of interest has been the development of sensor materials, especially those designed at the nanoscale. Sensors are structures that will respond in a recognizable way to the presence of something we wish to detect. There are sensors for temperature, water, light, sound, electricity, particular molecules, and specific biological targets such as bacteria, toxins, explosives, or DNA.

One way in which Hupp is trying to develop sensors is by using the properties of molecular recognition. He has made some rather complex and elegant molecules that he calls molecular metal squares, one of which is shown in Figure 5.1. These squares are designed to recognize particular target molecules also called *analytes* ("analyte" literally means that which we wish to analyze). By designing the molecular squares with particular geometries and patterns of molecular electron density, Hupp and his group have been able to perform a Cinderella-like act—the analyte foot fits into the molecular square shoe and other molecules with different sizes and shapes do not. Once the molecular square recognizes and captures the analyte molecule, we must be able to recognize that the capture has in fact taken place, which is usually done by shining light onto the square. The combination of square plus analyte absorbs energy from the light in a different color range (frequency or wavelength) than the square

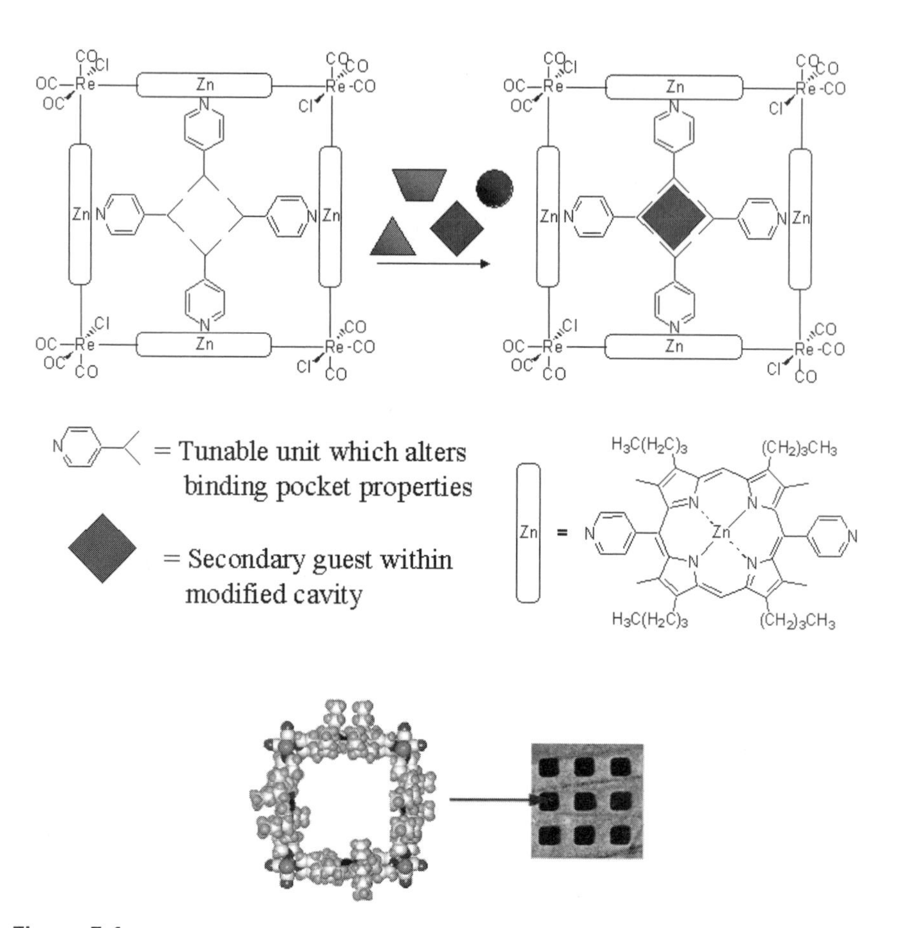

Figure 5.1
Synthetic chemical nanoscience—metal-trapping molecular squares.
Courtesy of the Hupp Group, Northwestern University.

without the analyte or the analyte alone. This means that if you monitor the sensor, it will change color in the presence of the analyte. These sensors are sensitive enough to detect fewer than 10 molecules of an analyte, so for high-precision tests you might not see the change with the naked eye, but it isn't hard to construct lab equipment that can see it. This allows the squares to be among the most sensitive sensors ever made.

Sensor technology is critical to the control and monitoring of the environment. The concept of sensors is not new—Humphrey Davy developed a miner's lamp that sensed the presence of gas in coal mines at the beginning of the 19th Century—but nanotechnology will make whole new classes of ultrasensitive sensors possible. In Chapter 7, we will discuss some of the nanoscale sensors that have been created, what their general properties are, and why this may be one of the first major commercial applications of nanotechnology.

NANOSCALE BIOSTRUCTURES . . . .

Sam Stupp teaches chemistry, materials science, and medicine at Northwestern University. Stupp grew up in Costa Rica and studied materials science and dental materials early in his career. He speaks several languages, is serious about literature and gourmet food, and is a scientific visionary. He heads an institute at Northwestern devoted to human repair, which means that one of his major research aims is the utilization of self-assembly and nanostructures to repair, rather than to remove or replace, parts of human bodies when they run into trouble. A major focus of his research, and a major focus of nanoscience generally, is so-called *nanoscale biostructures*. These structures, designed at the nanoscale, can mimic or affect a biological process or interact with a biological entity.

One example of a nanoscale biostructure is provided by a self-assembling "artificial bone," very recently developed in Stupp's group. Figure 5.2 shows the general notion: the molecules that make up the bone are held together by chemical bonds. These molecules, in turn, have interactions among them that are weaker than true bonds (more like those that create surface tension in water), but that hold the molecules together with each other in a particular shape, in this case a cylinder. The molecules in the bone are designed to occupy space in a particular way so that they will assemble spontaneously to form the desired shape, and, once assembled, so that they will be packed densely enough for the bone to be very strong. The structure of packed molecules can be made compatible with the human immune system by properly choosing the *head groups* of the molecule, the groups of atoms that ultimately form the outer shell of the artificial bone template. The outer shell is also designed so that natural bone begins to form around it like coral on a reef or gold on a piece of plated jewelry. This is key to human repair—allowing the body to fix bro-

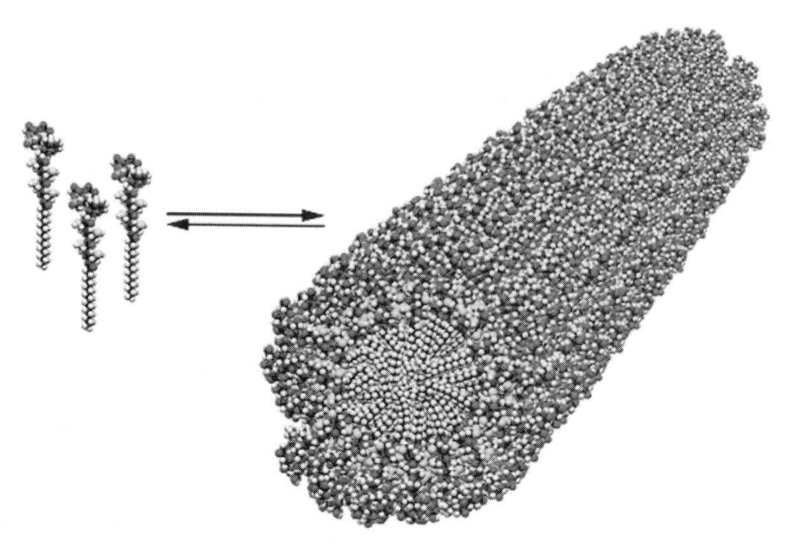

Figure 5.2
Self-assembled molecular template for an artificial bone. The long rod self-assembles from the small molecule components and natural bone tissue forms on the outside edge. *Courtesy of Stupp Group, Northwestern University.*

ken or damaged tissue naturally rather than replacing it with a steel or ceramic implant.

Because the biological realm is full of nanostructures, biomedical applications and biomedical investigations constitute a major part of the nanoscience landscape. Chapter 8 is devoted to a quick overview of some of the many areas of biomedical nanotechnology.

ENERGY CAPTURE, TRANSFORMATION, AND STORAGE

Michael Graetzel is a chemist at the University of Lausanne in Switzerland. He has curly hair, a shy and engaging smile, and a tremendous enthusiasm for what he does. Graetzel has devoted much of his career to the invention, study, and development of nanostructures for dealing with issues of energy—its capture, transformation, storage, and distribution. Because industrialized societies require massive amounts of energy, both in continuous supply to homes and businesses and in portable energy for gadgets and personal electronics, the area of energy management comprises one of the major domains of nanoscience.

Graetzel's first major contribution came in the development of something now called the Graetzel cell. In a *Graetzel cell,* a dye molecule is used to capture the energy from sunlight. The molecule absorbs the light, going into a higher energy state. In this high energy state, the molecule actually separates charge by passing an electron from the dye molecule to a nanoparticle of a white crystal called titanium dioxide, which may be familiar as the pigment material in white house paint. The separated charges (positive charge remaining on the dye molecule, negative charge shifted to the titanium dioxide nanoparticle) are then allowed to recombine using a set of electrochemical reactions. In this recombination, some of the energy that was originally captured from the sun by the molecule is released as electrical current passing through an external circuit. Originally, Graetzel cells were used to illuminate bathroom scales and Swiss watches, but they also exemplify a major worldwide effort for capturing sunlight to provide energy sources that are efficient, nonpolluting, safe, and inexpensive. Graetzel cells currently have efficiencies exceeding 7 percent and can be produced using silk screening techniques, which makes them cheaper to make than most traditional photovoltaic cells.

Chapter 9 deals with nanoparticle optics—the capture, control, emission, transmission, and manipulation of light. Because light is one of the most important sources of energy, this area of nanoscience and technology is crucial for dealing with the world's energy demands.

OPTICS

Ching Tang is a chemist at the Eastman Kodak Company in Rochester. Kodak's name has long been synonymous with one kind of optics, the kind that allows us to capture memories on film. Tang is soft spoken and charming, but his unassuming manner masks the tremendous creativity that he has shown throughout his scientific career. In 1987, Tang's group at Kodak was the first to demonstrate that organic molecules could be used to make light directly and efficiently from electricity. The area that Tang invented in that year has become known as *organic light-emitting* diodes (LEDs), and all signs are that it will be a major technology for illumination in areas from automobile dashboards to room lighting to computer screens.

Tang's original work was based on molecular considerations, rather than any nanoscale structures. More recent work throughout the world has demonstrated quite clearly that taking these LED structures down to the nanoscale produces major gains in efficiency, control, cost, and lifetime.

Using electricity to produce light, as is done in these LEDs, is essentially the inverse process of natural photosynthesis or of the Graetzel cell discussed earlier in this chapter. Electricity is used to produce light in the light-emitting structures, while light is captured to produce electricity in the photocells. Both are major areas of study in the development of nanoscience, and both are described in Chapter 9.

Experiment in the Palm of Your Hand: Lab-on-a-Chip

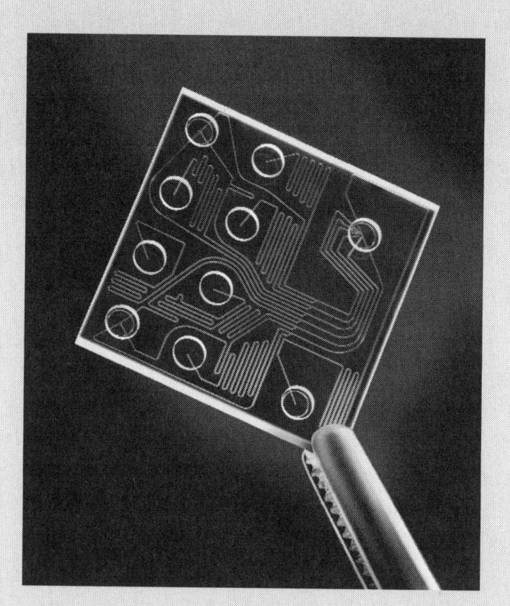

Figure 5.3
A lab-on-a-chip. *Courtesy of Agilent Technologies, Inc.*

If you read almost any scientific proposal, you will be struck by the amount of money required to pay for two simple but basic needs: space and people. Scientific instruments take up a fair amount of space, it's true, but most lab space requirements are taken up by passageways, desks, open surfaces, keyboards, monitors, hoods, emergency equipment, and other necessities for the humans who work there. If you could automate all human tasks in a laboratory and collapse all this space, you could make things much more compact and efficient. In some cases, you might be able to make them so compact and efficient that you could integrate the whole lab onto a microchip. By reducing these costs and overhead, you not only make research simpler, quicker, and cheaper, but also make it possible to conduct hundreds or even thousands of experiments at the same time.

That is the basic idea of an emerging technology appropriately named *lab-on-a-chip*. See Figure 5.3. At first glance, these tiny, automated laboratories look like their electronic brethren. They are usually created on silicon surfaces, and tiny cells are linked by microscopic or nanoscopic interconnects.

The difference is that, in lab-on-a-chip, interconnects don't all conduct electricity. Many of them channel fluid from tiny reservoirs implanted on the chips during fabrication. The functional cells are also different. In a microchip, these might be memories or logic gates, but in a lab-on-a-chip they are often mixing elements, reservoirs, and bio or chemical reactors.

Lab-on-a-chip fabrication is done using well-established silicon technologies including lithography and etching. Lab-on-a-chip differs from electronic chip making, however, because features must be designed in three dimensions instead of two. The reason for the three-dimensional design is that even though electricity may be able to flow through a planar wire, water can't flow through a flattened hose. Three-dimensional silicon manufacturing isn't as well understood as it is in two dimensions, and some of the plastics and other materials required to handle fluids are different from those required for handling electricity. These issues make lab-on-a-chip fabrication a lively area of engineering.

The other key technologies for creating a lab-on-a-chip are *microfluidics* and *nanofluidics*, the approaches to controlling the movement of fluids through channels at the microscale or nanoscale. When the volumes of fluid are this small, you can't always push fluids along using pumps or valves because you wouldn't have the precision required and because such small moving parts would be very hard to design and integrate. Instead, when very small fluid volumes are required, two techniques are used in current lab-on-a-chip devices: electrophoresis and electroosmosis. Both approaches work by applying a voltage difference along the channel in the direction the fluid should move. In electrophoresis, this voltage difference interacts with ions distributed throughout the fluid to be moved, pushing them along using Coulombic forces. When this approach is used, the ions in the fluid move at speeds inversely proportional to their mass, causing them to separate, with lighter particles moving faster and heavier ones moving slower. This separation by mass is why electrophoresis is useful in analyzing composition and is used in DNA analysis. *Electroosmosis*, on the other hand, works by having charges on the channel wall interact with a thin sheath of ions at the wall-fluid interface. This pushes the entire fluid column along at the same speed like a plug through a tube.

By using these approaches to move fluids among the mixing elements and reactors, it is possible to control interactions precisely, and the lab-on-a-chip has already become reality. Companies like Affymetrix (with their product GeneChip) and Agilent (with their product LabChip) make lab-on-a-chip devices for genetic analysis. It is hoped that these chips may develop to the point where they can be used for point-of-care applications so that a doctor can give a patient an immediate analysis of blood or any other samples that the doctor takes. They may also be used for drug delivery, particularly in cases where drugs need to be dispensed over a long period of time in response to changing body chemistry (as in diabetes, for example). In the more distant future, it is possible that lab-on-a-chip might serve as a framework for DNA computing since early experiments in that field operate in just microliters (a millionth of a liter) of solution but require large-scale analysis of results to be useful. Lab-on-a-chip might also be used for experiments in orbit on stations or shuttles where space, ahem, is truly at a premium.

Another Spin on Things: Quantum Computing

Shrinking microchip features into the nanoscale will offer us the ability to continue Moore's law for several more chip generations, but nanotechnology also offers some provocative possibilities for beating even that amazing trend. One of these possibilities is the notion of *quantum computing*, using the quantum properties of particles for doing computation. But what does that actually mean? One approach to quantum computing resides with our friend the all-purpose electron.

In addition to the properties that we are familiar with like mass and charge, electrons have a number of quantum properties, one of which is called *spin*. For our present purposes, what spin is doesn't actually matter. What does matter is that it has a value of either +1/2 or −1/2 and that it can be manipulated in interesting ways. For our discussion of quantum computing, it may be best to think of spin not as +1/2 and −1/2, but as computer-friendly binary values where +1/2 is a binary 1 and −1/2 is a binary 0. Assuming that we can control the value of the spin, we can now think of our electron as representing the smallest unit of digital information, 1 bit. Since quantum computers behave a little differently from conventional computers, we won't be content with just calling it a bit, however. The term of art is a *qubit*, not to be mistaken with the ancient measure "cubits," used by Noah to build his ark.

While it would be quite an accomplishment just to be able to represent a bit of information in a single electron, the laws of quantum mechanics reveal some other possibilities for qubits. No calculation can predict what value the spin has until you measure it, at which point the spin becomes fixed. Until that point, it behaves a little like 0 and a little like 1 and might be more readily thought of as both 0 and 1 at the same time, a personality disorder that quantum scientists call a *superposition of states* 0 and 1.

Why is this interesting? Although the spin isn't determined until you measure it, you can tweak it using light of specific frequencies. The light that you use and how you pulse and polarize it act as a program for your quantum computer. But the qubit's state (1 or 0) isn't determined while the program is running since you haven't measured it yet. Therefore, it effectively executes a given command as if it has both values, performing the two operations in parallel. Qubits can also be linked to one another so that the state of one affects the state of the others. This process is called *entanglement* and it is the key to making a computer with more than one qubit.

This ability to execute programs in parallel with all possible answers being represented is key to a number of interesting problems in computer science. Most cryptographic technology including RSA and DES, two of the most common encryption technologies on the Internet, is based on the idea that big numbers are very difficult to factor. A conventional computer, no matter how big and how fast, could take more time than there has been since the Big Bang to break codes that can be fairly easily created on a home PC. Quantum computing could change all that—by performing factoring in parallel it could break these codes quite quickly and easily. This is an example of an operation that is not just made faster by quantum computing, it is made possible. It is also a reason why quantum computing is considered so important since cryptography is key (as it were) to all digital security. Database searches are another algorithm that would benefit greatly from quantum computing.

But there are significant challenges in making quantum computers, and most of them relate to entanglement. The more electrons you have entangled, the greater the likelihood that some passing cosmic ray or other outside phenomenon will affect one of them and throw off your whole computation. This process is called *decoherence*. Currently, quantum computers with a handful of qubits have been made, but it seems unlikely that current approaches will be able to construct computers with more than 10 qubits. Adding a parity bit, an approach used in electronic computers to do error corrections when transmitting over uncertain media, is a possibility that might raise the number slightly. Since these test tube computers also tend to suffer decoherence after around 1,000 operations, there is clearly much work to be done. Still, the fact that this phenomenon has been demonstrated to be workable at all is very exciting, and much research is being done in this area.

One approach to solving this problem is to use electrons on nanodots instead of electrons on individual atoms as qubits. In this approach, nanowires are used to connect the nanodots and provide entanglement. This method provides an intriguing solution to the problem of controlling entanglement through the introduction of a physical connection, something that can't be done so easily between two atoms. It is also a grand illustration of the power of the nanoscale—the ability of bulk materials to be shaped into physical devices meets the quantum properties of the single electron to provide an all-new kind of computer.

Harnessing the Computer Inside Us: DNA Computing

The human body is, in many respects, an extremely efficient computer. One way in which it carries and processes data is through DNA and its complex biochemistry. Efforts to use the same approaches and techniques to perform general-purpose computing now comprise one of the most challenging areas of nanoscience.

DNA has several advantages for use as a computer component. For one thing, its data density is fantastic. DNA "bits" or base pairs (see discussion of DNA in Chapter 4) are packed onto DNA strands with a spacing of about one-third of a nanometer between pairs. This translates into approximately 100 megabits (millions of digital 1s and 0s) per inch or over a terabit (trillions of bits) per square inch. This density alone would be enough to get hard-drive makers' attention (current drive densities are much lower in density than this), but DNA can also be packed efficiently in three dimensions, making its data storage capabilities even more remarkable. Using DNA might never be efficient for high-speed, random-access applications of the type used in current hard-disk technology, but its potential as an approach to data archiving (the task normally performed by magnetic tape drives today) is enormous.

Double-stranded DNA is also naturally redundant. Single DNA strands bind with their natural complements (a strand that has exactly the opposite ordering of base pairs) in a process called *hybridization*. Hybridized DNA forms the familiar double-helix or double-spiral that is the usual depiction of DNA. Hybridization means that DNA has built-in fault-tolerance since a given bit of data actually resides on two strands. Disk makers call the hard-drive equivalent of this *mirroring*. Despite these advantages of density and redundancy, most natural processes for reading and copying DNA still have an error rate of more than 1,000 times that of state-of-the-art magnetic storage. Hopefully these problems will be overcome so that DNA will have a place in next-generation data storage.

DNA has other computing applications. By applying the natural processes that the body uses to read and write genetic information, researchers have been able to perform computations using DNA. In particular, a special-purpose computer (in computer science terms a "finite automaton") has already been demonstrated using DNA.

Perhaps the easiest example of a finite automaton is an elevator in a two-story building. It has two states: on the first floor and on the second floor. The elevator is always in one of these two states or moving (transitioning) between the two. It is simplest to think of a two-floor elevator, but the problem can be generalized for any number of floors. The elevator can accept two possible inputs: a request for the first floor or a request for the second floor. The elevator knows what to do at any given time based upon its current state and the current input. For example, consider the transition rules shown in Table 5.1.

Table 5.1
Transition Rules

State	Input	What To Do
On first floor	First floor request	Nothing: sit still
On first floor	Second floor request	Go to second floor
On second floor	First floor request	Go to first floor
On second floor	Second floor request	Nothing: sit still

These are called transition rules since they govern the transition from state to state. If you can code states, transition rules, and inputs, you can construct computers of this type. And, if scaled big enough, they have uses such as parsing text and doing pattern matching, applications that can be useful for everything from cryptography to speech recognition.

Some current DNA computers work using enzymes (restriction nuclease and ligase for the biochemists out there) as hardware, a double strand of DNA as input data, and a few short DNA molecules as transition rules or software. The input DNA codes the initial state (which floor it starts on) as a base-pair sequence and then the input data (floor requests) as additional base-pair sequences in order. After the initial state has been decoded (on first floor, for example), the machine proceeds to function by cutting or cleaving the DNA after the initial state. When DNA is cleaved, it exposes a sticky end, and the composition of the sticky end (which controls what it will bind to) depends on the next sequence of base pairs after the cutting site, which is the next input instruction (second floor request). In our device, only one of the possible software molecules can adhere to the sticky end (in this example, that software molecule represents the transition rule, go to second floor), and each of these software molecules is of a different length than the others. The next time the DNA is cleaved, the length of the software molecule controls the position of the cleavage site and therefore of the newly exposed sticky end. In this way, the state is controlled, and the cycle of "cut last input and bind to software molecule" continues until the input is used up or a special terminator sequence is cleaved, resulting in the generation of a special, easily detectable output molecule. This output molecule represents the final state of the machine.

These experimental DNA machines have run with speeds on the order of one billion transitions per second with an error rate of less than 0.2 percent. This speed is impressive—on a par with some PCs—but the error rate is still much higher than an electronic computer's. This DNA computer does have a few benefits over its electronic cousins, though. For one thing, the process uses only one-ten-billionth of a watt of power as compared to an electronic microprocessor, which consumes tens of watts. The fact that electronic processors use so much power has been an obstacle to their development because power is dissipated as heat and processors can get so hot that they damage themselves. It seems that this won't happen with DNA.

Low power consumption is nice, but the killer application of nano-scale processing (quantum computing, swarm computing, etc.) is that it can operate in parallel. Once you have designed a program in DNA format, you can mix in as many inputs as you like and have them all processed at the same time. Already, the first experiments have had one trillion processes operating in parallel, as opposed to the largest scale supercomputers, which have only a few hundred processors and certainly don't fit in a test tube.

Like quantum computing, the applications of DNA computing may remain special-purpose, but for DNA computing practical applications may come sooner. DNA synthesis techniques for making arbitrary sequences are becoming simpler, and custom DNA can already be ordered on a reasonable budget. Once this becomes an even more rapid and inexpensive process, unlocking the power of very large scale parallel processing may be at hand.

MAGNETS

Chris Murray is a young, inventive, and accomplished nanoscientist who works at IBM Watson Laboratories in New York. Since IBM has long been at the forefront of computing and data storage, it is not surprising that Chris is one of the world leaders in the area of ultra-high-density magnetic storage. The advent of magnetic disk technology, based on an important property of certain magnetic materials called giant magneto resistance, substantially brought down the price of computer memory and enormously increased the efficiency of computers. Murray is working to bring these magnetic storage elements to their nanoscale limit.

In his lab, Murray prepares individual quantum dots of magnetic materials. When these dots become too small, they cannot maintain their magnetic properties because thermal energies erase the magnetic signature. For this reason, Murray has focused on preparing dots that are just small enough (the technical term is "above the super paramagnetic limit") to retain their magnetism and, therefore, to retain the memory of the magnetic field that wrote them. Murray's work in preparing, stabilizing, measuring, and understanding such quantum dots is an example of *nanoscale magnetic structures*. Magnetic nanostructures are discussed in Chapter 9.

FABRICATION

Mark Reed lives in rural Connecticut, has a magnetic smile, a ready laugh, and has already pioneered nanoscience in several guises. While at Texas Instruments he was one of the originators of quantum dot circuits. He now teaches electrical engineering at Yale, where he has been a leader in building molecular electronics circuitry. He uses a number of techniques including electron-beam lithography, molecular self-assembly, and scanning probe instrumentation to build nanostructures and then measure their properties. The continued study and enhancement of nanofabrication techniques is a crucial activity in nanoscience and nanotechnology—after all, a structure that cannot be built is not very useful. Reed's work has been central to the whole area of molecular electronics and has other applications as well. We've already seen structures created using dip pen nanolithography (the nano-graffiti) and scanning probe techniques (the abacus). Nanofabricators are working to make even more complex structures and to make them much more quickly and efficiently.

Because nanofabrication is so important, most nanoscience centers are full of researchers who nanofabricate. For instance, in the Northwestern center, the cast includes Joe Hupp, Rick Van Duyne, Sam Stupp, Chad Mirkin, Mike Wasielewski (all of whom we have visited on the tour), and Teri Odom, the newest faculty member. Teri worked with nanoscience pioneers Charles Lieber and George Whitesides, and her work (like Reed's) does nanofabrication using both hard (metals, nanotubes, semiconductors) and soft (molecules) materials.

Fabrication is an omnipresent theme of nanostructures, and it will appear in all the subsequent chapters of this book.

ELECTRONICS . . . .

Mark Hersam is in the Materials Science Department at Northwestern. Hersam is a junior faculty member, whose independent career started only two years ago, but he has already begun to make a substantial name for himself whenever he can keep off the golf course. Starting with his doctoral work at the University of Illinois with Joe Lyding, Hersam has been a major developer of new methods for preparing and measuring nanostructures with important, unusual, and promising electronic properties.

Hersam's work is devoted to the subject of molecular electronics, the electronic properties of individual molecules. In ordinary circuits, this measurement could be done with a voltmeter, ammeter, or oscilloscope, but it is impossible to put a pair of alligator clips on a single molecule. This means that making a measurement at the nanoscale that is trivial at the macroscale can be extremely complex. Hersam approaches the problem by preparing a single crystal of silicon, one face of which is coated with hydrogen atoms. He then uses an STM to blast off an individual hydrogen atom and brings a molecule from the surrounding vapor to where that hydrogen atom was located. The new molecule binds where the hydrogen atom came off. The structure now looks like a flat plane of hydrogen atoms with one lone molecule sitting in the middle. This precise placement of single molecules is called *controlled feedback lithography*. Having placed the single molecule on the silicon, Hersam can then use scanning-probe methods to measure the current through that molecule as well as the motions that the molecule makes and how the current passing through the molecule affects this motion. While he makes these measurements, the silicon holds the molecule in place.

The notion of current passing through single molecules is one of the founding ideas of molecular electronics, one of the most important parts of nanoscale electronics, which we'll look at more in Chapter 9. More generally, application of nanostructures in electronics is one of the most vibrant and challenging aspects of nanoscience. Indeed, charge transport on the molecular or nanoscale level draws on a whole set of new concepts that challenge our understanding of electronics.

ELECTRONICS AGAIN

Bell Laboratories/Lucent Technologies has been the most successful industrial laboratory in history. Its staff members invented the laser and the transistor, and its contributions to modern physics and technology in nearly all guises have been remarkable. Zhenan Bao did her graduate work at the University of Chicago and has been at Bell Labs for her entire professional career. She is a young woman of tremendous accomplishment, charm, and creativity.

Zhenan is an organic chemist by training and her work at Bell Labs has focused on applications of organic molecules to nanoscience and nanotechnology. She has been at the center of tremendous advances at Bell Labs. Working with colleagues, such as Howard Katz and Ananth Dodabalapur, she has built a series of organic-based devices that use organic molecules to perform the sorts of functions more generally associated with silicon technology. Major advances have included inexpensive organic transistor chips to act as tags for identification of products, parcels, and postage. This work makes the promise of molecular electronics more than just a smaller version of conventional electronics, but a field where entirely new things are possible.

MODELING

George Schatz is a Professor of Chemistry at Northwestern University. He grew up near Watertown, New York, where his mother taught in a one-room schoolhouse and he lettered in several sports as a high school student. George is a theoretical chemist; most of his research deals with the behavior of molecules interacting either with other molecules or with surfaces.

One of Schatz's most powerful recent research efforts has been in the area of nanoscale optics. In particular, he has been concerned with predicting how the size, shape, and surroundings of nanoparticles change their optical properties—this is the phenomenon that underlies the change in color of stained glass windows with the size of the gold nanoparticles that they contain. George has generalized a model for the optical properties of nanoparticles that was invented a century ago by a German physicist named Mie and has developed very powerful computer programs for relating precisely how the size, shape, composition, and solvent environment of nanoparticles determine their color. These models permit tailoring of nanostructures so that they will respond to particular colors of light. This development has major implications for the design and application of nanostructures in such areas as light energy capture from the sun, signaling capabilities, or therapeutic agents.

Modeling and theory are at the heart of our understanding and design of nanostructures and, more generally, of the entire scientific universe. If general models can be created to represent a particular nanoscale phenomenon, it is possible to determine what nanostructures might exhibit that characteristic more strongly or more usefully. Modeling allows us to speak with confidence when we say that nanotubes are not only the strongest material yet made but also the strongest material our current understanding of science finds possible. When you can engineer at the molecular level, it is important to understand not only what you've done, but also what you could do if anything were possible, since with sufficiently advanced fabrication techniques pretty much anything that is stable can in fact be made. Even though there is not a specific chapter in this book devoted to modeling, it is a critical enabler for all of nanoscience because it provides the basis for design and understanding.

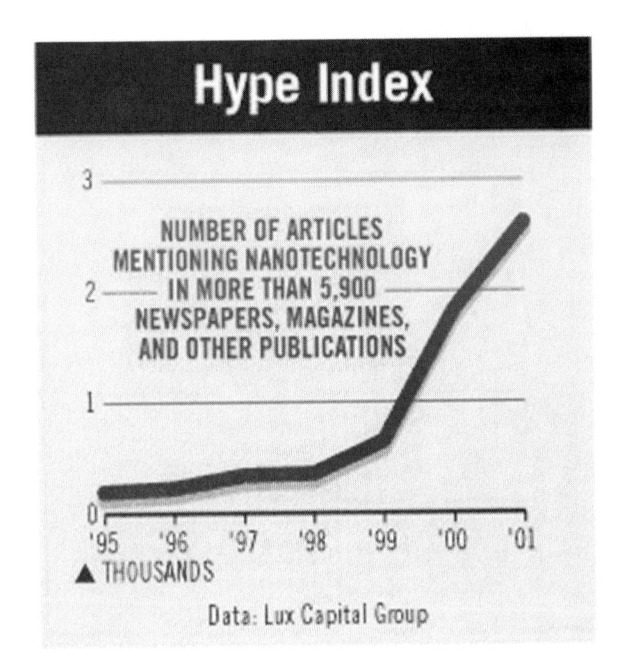

Figure 5.4
The nanotechnology hype index. *Courtesy of Lux Capital.*

Nano concepts, nanoscience, and nanotechnology are growing at a tremendous rate. Figure 5.4 shows the number of scientific papers published annually using the word "nanoscience"—notice that it goes from zero 10 years ago to well over twenty-five hundred and climbing papers per year, and it is growing at a rate similar to Moore's law. The very abbreviated grand tour of topics we have just completed suggests the excitement and the promise of nanoscience. Now it's time to dive into some detail.

6 Smart Materials

Nano-scale science and engineering most likely will produce the strategic technology breakthroughs of tomorrow. Our ability to work at the molecular level, atom by atom, to create something new, something we can manufacture from the "bottom up," opens up huge vistas for many of us....This technology may be the key that turns the dream of space exploration into reality.

David Swain
Senior Vice President
of Engineering & Technology,
The Boeing Company

In this chapter...

Suppose that corrosion processes could be effectively stopped so that bridges and railroads could be maintained at a fraction of the current cost. Suppose that stain preventers could be incorporated permanently into clothing so that spilled soup would no longer mean a trip to the dry cleaner. Suppose that automobile windshields did not get wet so that no ice would ever form on them and rain could not impede visibility. Suppose that bathroom tiles and hospital sheets could be developed that would self-clean, killing any bacteria or virus that settles on them. Suppose that automobile windows could automatically adjust their reflections to the prevailing sun so that a car parked in a Phoenix afternoon would remain at a civilized temperature. Suppose that a rip in a fabric or a puncture in a tire could immediately and automatically repair itself. All these things are possible, and some are already a reality. They come from the use of smart materials.

What makes a material "smart" is that it incorporates in its design a capability to perform several specific tasks—in nanotechnology, that design is done at the molecular level. Smart materials can function either statically or dynamically, which means that some of them always behave the same way and some react to outside stimuli and actively change their properties. For example, Teflon is a smart material because it is engineered to have essentially no stick, and it is a static smart material because it isn't designed to react to external forces. Concept stealth fighters, however, are coated with a dynamic smart polymer material that changes its color and electromagnetic signature in response to outside conditions and pilot instructions, providing the ultimate in camouflage. The design of smart materials is a major technical challenge and a major economic opportunity for nanotechnology.

Nearly all biological structures are smart materials. A remarkable example is human skin. Skin is permeable to certain substances such as water and dissolved ions. It acts as a sensor to heat, to touch, and to sound. It is self-renewing. It also acts as a barrier to air from the outside and to biological fluids from the inside. Skin is a multicomponent dynamic smart material, showing some of the features that nanotechnology attempts to design into synthetic structures.

Smart materials are not necessarily nanoscale phenomena—Teflon pans are designed at the macroscale, as is antifouling paint on boats.

But the capability for nanoscale design provides for far richer and smarter materials than can be realized with macroscale components. The ability to work at the nanoscale, "the ultimate level of finesse," allows us to create materials that can leverage molecular properties, macroscopic properties of bulk materials, and even biological processes into the construction of smart materials. It is clear that smart materials comprise a very large range of structures and activities, and many are becoming the focus of substantial attention. Let's look at some of them.

SELF-HEALING STRUCTURES . . . .

When one thinks of healing, the first example that comes to mind is the body's self-repair of cuts and bruises. When a blood vessel is ruptured, platelets come together to form a clot that stops the bleeding. The process of healing can then continue until the blood vessel is fully repaired. This is a very complex operation involving several blood components and cell growth, but the process can be simplified and applied to artificial structures.

The simplest *self-healing structures* respond to local breaks in a continuous fabric by repairing them. Self-healing automobile tires are a classic example in which a new polymer is formed to bridge over punctures in the initial polymeric tire structure. Clearly, these tires repair macroscopic rather than nanoscopic holes, but the principle of self-healing is there.

An even more common example of a simple self-healing structure is cooking oil in a frying pan. Scraping a spatula across the bottom of the pan will temporarily remove oil from the line that follows the moving spatula. The clear line will not last long, however, because the cohesion of the molecules in the film with one another will cause the oil to flow back together, seamlessly covering the dry scraped trail with oil.

One true nanoscale example of self-healing occurs in typical biological membranes. In the simplest picture, these membranes are held together by molecules that are shaped like balloons with long strings tied around them. The fat nanoscale balloon part of the molecule is either charged or strongly polar, so it dissolves in water. The long

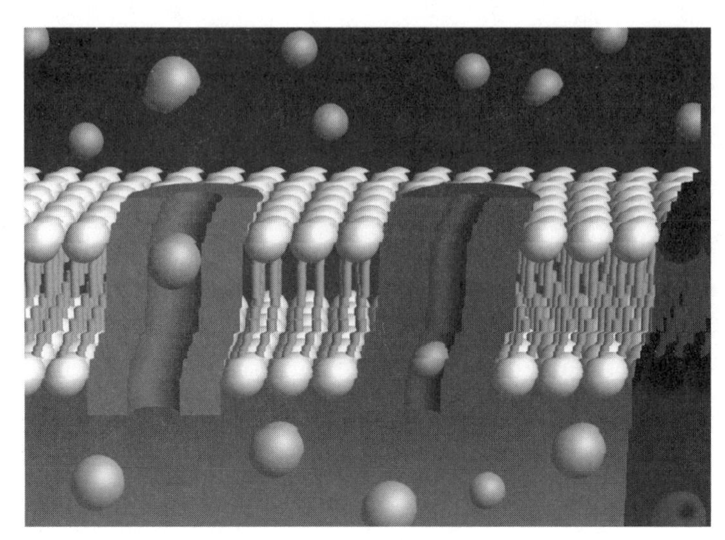

Figure 6.1
A computer-generated model of a portion of cell membrane. The light balloons are hydrophilic, and the dark thin strands are hydrophobic. The cylinder structures are channels for moving ions through the membrane.
From General Chemistry, *8/e, by Petrucci/Howard, © Pearson Education, Inc. Reprinted by permission of Pearson Education, Inc., Upper Saddle River, NJ.*

nanoscale string is uncharged and nonpolar, so it won't dissolve in water. A sketch of this structure is shown in Figure 6.1. Because the polar head groups (the balloons) are soluble in water, they tend to group together. Similarly, the nonpolar, greasy string parts cluster together because they have more chemical attraction to each other than to the water. The result is that the simplest picture of a biological membrane shows polar, hydrophilic (water-loving) head groups poking out into the water structure and nonpolar, so-called hydrophobic (water-hating), hydrocarbon tail groups in the middle of the membrane. The thickness of the membrane is generally around 1 to 20 nanometers.

If a hole were to be poked into this membrane, the balloon-shaped molecules would immediately move to fill that hole. The biological membrane would repair itself even if the hole were very large, unless another object occupied the hole (sometimes other nanostructures sit on the surface of the membrane and block repair—these can include ion channels that allow ions to enter cells and the apparatus for photosynthesis). Part of the secret behind the remarkable stability of biological cell membranes is that they are self-repairing on the nanoscale. Self-healing properties can be engineered into a variety of materials and are now seeing use in engineering plastics.

RECOGNITION

One way in which a material can be smart is to react only under certain conditions. Another way materials can be smart is to permit separation of substances to which the material is exposed. Materials can also be smart based on molecular recognition capabilities that permit them to respond to particular chemical or electromagnetic perturbations or stimuli.

We mentioned nanotube and nanowire growth from single crystals in Chapter 4. This is a matter of recognition—the different components of the nanowire first recognize the seed crystal, which is placed in their midst, and then recognize each other. In this way, the solid nanowire expands into a bathing solution, in exactly the same way that an icicle expands by freezing water on its outside, salt crystals expand in a saturated solution of salt, or rock candy crystallizes from a sugar solution. A particularly ingenious idea used in several laboratories (such as those of Charles Lieber at Harvard or Peidong Yang at Berkeley) utilizes seed crystals that cause the growing structures to choose the particular morphology (shape) of a long thin wire (Figure 4.7), rather than the more characteristic morphologies of block crystalline structures. This is an example of molecular recognition being used to create particular nanoscale structures. It is also an example of classic crystal or fiber growing taken to the ultimate level of finesse.

Another example is found in the biological cell. We've just discussed the bilayer membranes, with hydrophilic outsides and hydrophobic insides that encapsulate biological cells. To move things through the membrane, biology developed so-called channels, which are effectively just water-filled pipes. These pipes are nanoscale in cross section, and permit nutrients, wastes, and other critical ingredients to pass between the cell and its environment. Reza Ghadiri and his group at Scripps Institute have used self-assembled structures of ring-shaped molecules called cyclic small peptides to construct artificial channels. These small peptides stack on top of one another into an arrangement that looks like a nanoscale stack of tires and makes an artificial channel. Such artificial channels can be introduced into the cell membrane, causing things to leak in and out of the cell with great rapidity. Some possible medical applications in causing the death of cancer cells can be imagined for this form of self-assembled smart material.

The natural channels, and the artificial channels that Ghadiri and others make, work by molecular recognition in two different areas: the components of the channel recognize one another, and the assembled channel recognizes its external bilayer membrane environment so that it can insert itself. The combination of molecular recognition and assembly can result in materials that are smart on many levels and are only possible on the nanoscale.

SEPARATION

Separating a mixture of molecules or materials into its components is an important process both in biological systems and in the chemical, food, waste treatment, and pharmaceutical industries. In nature, selective reactivity is a very common form of separation; as a person digests food, the digestive system separates the sugars that have nutritional value from the food that does not. In industry, separations are usually performed by a physical process that permits one component to be segregated directly from the others as is done with distilling columns in oil refineries.

Applications of nanostructures in separations can be as simple as the cellophane that is used in packaging, which permits small molecules to pass through its nanoscopic pores while blocking other larger species. Precisely the same principle applies in dialysis, an important but extremely taxing procedure by which sufferers from kidney disease literally have impurities washed out of their blood. In current dialysis technology, blood must be pumped out of the body, filtered through a dialysis machine, and then pumped back into the body. There have been advances in dialysis membrane technology, but this technology remains a major challenge to the chemical industry.

Gore-Tex fabric is a smart material with two functions: it allows water vapor to escape (so that the wearer does not feel clammy), and it keeps the liquid water out (so that the wearer does not get wet). In this material, a polymeric sheet is pierced by tiny holes. The sheet is a fluorinated carbon, like Teflon. The holes are nearly nanoscale in size and will allow vapor molecules or small clusters of molecules to pass, while preventing liquid from passing.

Nanostructures are used in separations science in several ways. One obvious scheme involves molecular recognition, developing a particular molecular site that can specifically bind to a desired molecular target within a mixture. For example, the large molecule called EDTA (ethylenediaminetetraacetic acid) has four special acidic oxygen sites on the ends of flexible arms. EDTA can be used to capture various metal ions from solution. Generalizations of the EDTA concept involve so-called *siderophores,* molecules that are specifically designed to use their flexible arms, studded with particular charge species containing nitrogens, sulfurs, and oxygens, to capture desired metal ions. Siderophores look and act a bit like nanoscale octopuses: their molecular arms are like tentacles and their charge species act like suction cups to capture metal ions. Ken Raymond's group at the University of California at Berkeley has developed siderophores to capture particular metal ions that might be toxic. This approach could be used to capture and eliminate toxins such as arsenic or even lead in the body; consequently, it can be of great value for workplace safety and water treatment and is a beautiful form of designed molecular nanostructures.

The most common way to separate things is actually to pass them through holes like powder in a sieve—both polymers and crystals with small holes work very well in separations. This topic is generally referred to as *ultrafiltration* or *nanofiltration* and can be of tremendous economic value. The Air Products Corporation derives much of its income from using chemical ways to separate oxygen and nitrogen from air and then selling those pure gasses for industrial, chemical, and medical uses. Nanofiltration has other applications ranging from water purification to removal of toxins from waste streams. All these applications can be facilitated by the construction of nanoscale porous structures, and the development of such porous structures by industry leaders such as the Dow Chemical Company is a very significant commercial application of nanotechnology.

CATALYSTS

A *catalyst* is something that makes a chemical reaction go more rapidly. In the body, *enzymes* are the most common catalysts; they are the protein molecules that specifically speed up certain chemical reactions. For example, ptyalin in saliva facilitates the breakdown of starch into simple sugars, which is why bread tastes sweet if you hold it in your mouth for a while. In the fields of preparative synthetic chemistry and chemical engineering, catalysis is one of the greatest economic contributors because it can be used for applications such as oil refining. Starting with crude oil, catalysts are used to make gasoline and jet fuel and various hydrocarbon molecules, which in turn can be used to make plastics and petrochemicals.

One approach to increasing chemical reactivity is to take advantage of the fact that reactivity is related to surface area. If something has more surface area, there are more places for other chemical agents to bind, interact, and react. As particle size decreases, surface area increases, assuming the total mass of the material stays the same, just as in our cube of gold from Chapter 2. Therefore, as their particle size drops into the nanoscale, materials have the maximum possible surface area and therefore the maximum possible reactivity, which is the aim of a catalyst. This approach is used by companies such as Nanomat, which make materials with nanoscale grains for industrial applications.

Nanoparticles can sometimes be used as nanoscale catalysts, but this represents only the ultimate scaling down of the existing technology of using finely divided powders as catalysts. However, nanotechnology also offers some entirely new opportunities, especially when combined with separation techniques. In particular, starting with work by the Mobil Corporation, there has been tremendous interest in using structures called zeolites for directed catalysis because they produce petroleum more efficiently and can be used to select particular desired molecular products from the broader distribution of petroleum components.

Zeolites are often referred to as molecular sieves because their physical shapes allow them to sift materials. In structure they look like nanoscopic galleries or chambers interconnected by nanoscopic tunnels or pores, all dug out of a solid oxide. (See Figure 6.2.)

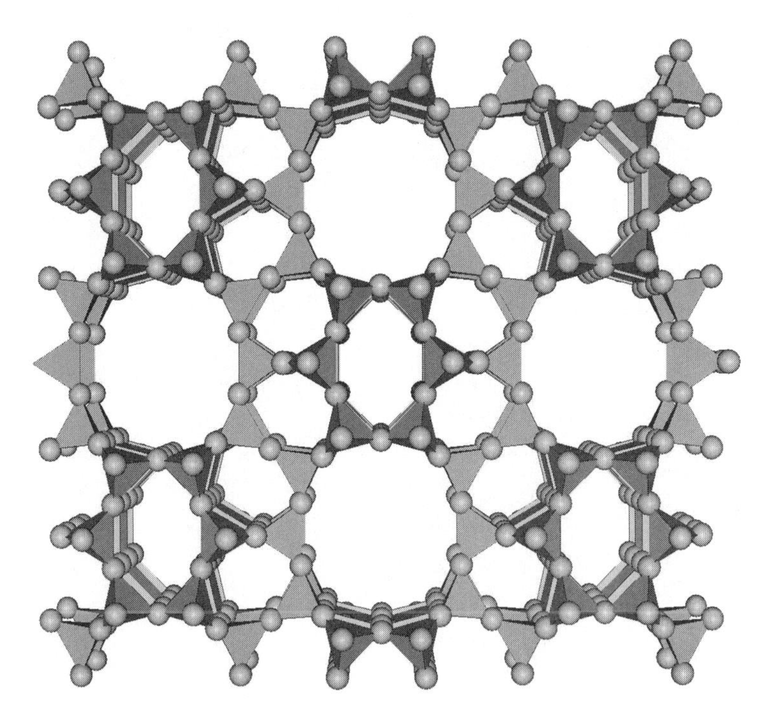

Figure 6.2
A chemical model of a complex zeolite structure. Notice the differently sized holes that represent channels and galleries. *Image courtesy of Geoffrey Price, University of Tulsa.*

In zeolite catalysis, particles of a catalyst are found within the galleries of zeolite. The combination of locally controlled reactivity by these catalyst particles with the physical constraints of gallery and pore size favors hydrocarbons of a particular shape and composition. This process of catalysis by design, as opposed to the random catalysis that was more common in previous generations, has resulted in more efficient use of feedstocks, less waste, and lower cost.

Zeolites are actually quite common; they are often used as domestic water softeners, where they facilitate the exchange of calcium ions by sodium ions and reduce water hardness. There are hundreds of different zeolite structures, both synthetic and artificial. The special nanopore structure of zeolites is the secret of their catalytic capabilities, and they represent one of the first broad-scale, highly profitable applications of nanotechnology.

HETEROGENEOUS NANOSTRUCTURES AND COMPOSITES . . . .

Some nanostructures are homogeneous—the examples of gold nanodots in stained glass or titanium dioxide nanodots for battery applications come to mind. There are also many heterogeneous nanostructures and nanocomposites. Heterogeneous means, in this case, that the material is not the same physically throughout its bulk.

One simple but lovely example is offered by core/shell nanoparticles. Many groups worldwide have made these structures for several purposes. They usually resemble nanoscale gumballs, with an inner core of one material and a thin outer shell of a second material. Chad Mirkin's group has used silver-core, gold-coated nanodots in DNA detection, and Moungi Bawendi's group at MIT pioneered the study of core/shell semiconductor nanoparticles.

A very simple class of heterogeneous nanostructures includes reinforced structural materials. Think of reinforced concrete, which is just ordinary concrete poured over a framework of metal rods called rebar. If the concrete is replaced by a plastic and the rebar is replaced by strong, firm, rigid nanotubes, the result is a nanostructured composite material with great fracture strength. These are smart materials, in that they are structurally designed for a particular application, and it is certain that nanotechnology will produce a host of them with unprecedented strength and versatility.

There is a huge market for these innovations. Boeing and AirBus are now spending some $50 billion on next-generation aircraft like the 550-seat A380 super jumbo and the nearly supersonic *Sonic Cruiser*. Both aircraft use carbon fiber materials extensively to keep aircraft light enough and strong enough to fly. Carbon fiber is a good material, but it cannot match engineering properties of *nanocomposites*. Nanocomposites are stronger, creating greater durability and requiring less total material. They are also lighter, allowing faster flight and greater fuel efficiency for cheaper operation. Although they will have application throughout the economy, aircraft engineering is a huge industry that could immediately profit from smart materials breakthroughs.

As we learn more about nanostructures and nanofabrication, preparing complex materials with several properties becomes possible. For aircraft, it might be interesting to combine self-healing with nanocomposite physical properties. This configuration might allow an aircraft to recover from the sort of fuel tank damage that downed Air France's *Concorde* flight 4590 or even the space shuttle *Challenger*. On a more personal note, it is promising to combine separation, recognition, and control of electronic structures to make a woven fabric of mixed smart fibers. These fabrics can change colors when activated by a battery and change porosity after sensing some particular molecule or detecting a certain level of heat and/or humidity. A shirt made of this fabric could change from a yellow open weave on a hot afternoon to dark blue, warm, and woolly in the cooler evening. This capability could be very convenient, but it is probably best to have pants from the same material or it could be almost impossible to accessorize.

ENCAPSULATION

Because hollow structures can be made at the nanoscale (like zeolites), it is possible to make *encapsulated nanomaterials*. In biology, these are actually rather common. For example, metal ions such as zinc or copper are normally not isolated within the cells of the body. Rather, they are surrounded by small proteins, some of which are called chaperones. Tom O'Halloran's group at Northwestern has shown quite clearly that copper is moved across the cell from site to site surrounded by small protein *chaperones*. These are smart structures in that they recognize the presence of a copper ion, encapsulate it, move it to another place, and hand it over. This is the same sort of process we discussed with siderophores, our nanoscale octopuses.

It is also possible to build glass houses around small structures (throwing stones is not a problem at the nanoscale). Some such structures are encapsulated enzymes, in which glass nanospheres, with holes in them, are used to enclose an enzyme. The enzyme may have some function, such as binding oxygen or moving electrons onto some molecular target. Because the enzyme is encapsulated partially in the glass, it is protected from certain chemical or physical attacks. Because the enzyme is partly exposed, it can still continue to do its work. Joe Hupp's group is one of several that have worked extensively on these encapsulated enzyme structures.

A more down-to-earth application has been pursued by David Avnir, his group at the Hebrew University in Jerusalem, and the start-up he cofounded, Sol Gel Technologies. They have used these nanoscale glassy spheres, produced by a process called sol-gel glass forming, to encapsulate suntan oil. If you have ever greased yourself up before going to the beach, you realize how unpleasant suntan oil can be for your eyes, your clothing, your comfort, and the aesthetic value of sunbathers in general. By placing the creams within nanoscale glass structures, Avnir has been able to avoid many of these greasy, oily sensations. Moreover, the organic molecules within the glass cannot interact chemically with the body. This separation might be important because some of the molecules found in ordinary suntan oil not only promote tanning but also can interact chemically with the skin and cause itching and rashes. Suntan oils are also quite possibly carcinogenic. Encapsulated suntan lotions avoid these problems and are a beachfront example of smart materials.

CONSUMER GOODS

The usual arguments for the importance of nanostructures and nanoscience are made in terms of electronics, and we began this way in the first two chapters. But modern society also depends on consumer goods, to make life easy and pleasant. We've already mentioned Gore-Tex, which certainly improves the life of any camper, and Teflon, which does the same for any cook. There are dozens of other examples of nanoscale-designed smart materials that improve the life of the consumer.

We've already started discussing the potential impact of nanotechnology on our wardrobes, and the first few products of this type are already shipping. Some of them come from a company called Nano-Tex, which uses nanoscale structures to change the physical properties of clothing. Its Nano-Dry product purports to combine the strength of synthetic material such as nylon with the comfort of natural open-weaves by carrying away perspiration and moisture. It does this by wrapping strong, durable, synthetic core fibers with specially made absorbent fibers to produce what could be called a smart heterogeneous composite nanomaterial, though we agree that calling it Nano-Dry is probably catchier. (See Figure 6.3.)

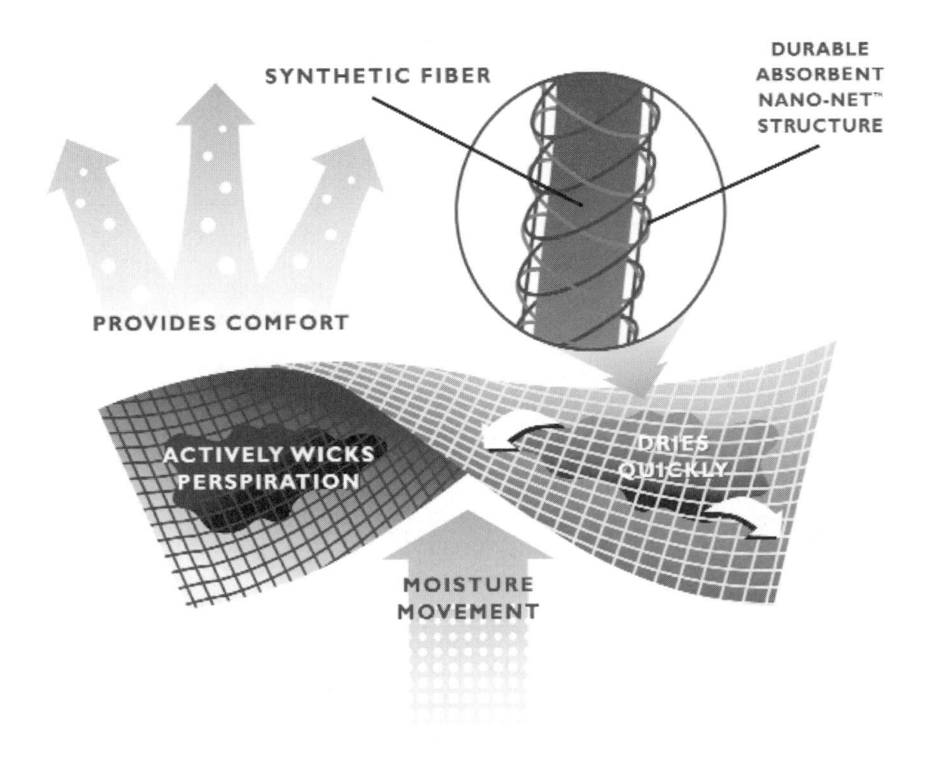

Figure 6.3
Nano-Dry. *Image courtesy of Nano-Tex, LLC.*

We have seen in this chapter some of the many examples of how nanoscale design can result in improved behavior of medical, physical, commercial, and consumer products. Designing at the nanoscale can lead to materials that have remarkable physical and chemical properties, materials that can respond dynamically to stresses placed upon them, and materials that can protect us and the structures in our environment. The number of features that can be designed into a material will grow daily along with our knowledge of nanostructures.

7 Sensors

In this chapter...

Imagine supermarket packaging that could tell instantaneously if the food inside had been exposed to heat or had begun to spoil. Imagine a tiny foolproof automatic indicator that could constantly monitor a home for dangerous chemicals such as gas leaks, carbon monoxide, or ozone. Imagine a simple swab test that would instantly tell whether a patient has strep, diabetes, various genetic diseases, the flu, or an anemia. Imagine a simple wire that could be inserted in the ground to tell a home gardener where the soil is best for planting cucumbers. Imagine a foolproof (and it would need to be) explosives test so that you wouldn't need to remove all your clothes at the airport. All these possibilities are potential applications of nanoscale sensors. Sensors are structures that indicate the presence of particular molecules or biological structures, as well as the amounts that are present. Sensors are already present throughout our society, but the best sensors will be made from nanostructures and, for starters, should revolutionize much of the medical care and the food packaging industries.

Sensors are newer than you might think. Most people's earliest memory of the word is from Mr. Spock in the original *Star Trek* series. In fact, the word isn't much older that that. It was first used in an article in the *New Scientist* magazine in 1958 and is defined by the *Oxford English Dictionary* as "a device giving a signal for the detection or measurement of a physical [or chemical] property to which it responds." Actually, the dictionary definition does not include the information in the brackets, but the application of sensors to the detection of molecular structures is perhaps the most promising and important area of sensing activity.

NATURAL NANOSCALE SENSORS . . . .

As is true with so much of the rest of nanoscience and nanotechnology, examples of sensors at the nanoscale are very widespread in biology. Sensors are crucial to communications, and communication with other organisms is one of the central characteristics of life. Signals come in a variety of formats including molecules, sound, smell, and touch, and they also can come in electromagnetic forms such as heat and light. The ability to detect these signals is both desirable, as in a fragrant perfume, and necessary, as in the detection of mercaptans, which are the nasty-smelling sulfur-containing substances that are added to the natural gas piped into houses.

The exquisite nanosensors in the nasal bulb of some animals, particularly dogs, is crucial to their survival and to some of the ways in which they help people. The fundamental mechanism behind a dog's sense of smell, or behind the sex attractants (pheromones) that are crucial in much of the insect world, is molecular recognition. Complementary shapes within the sensing structure of the dog's nose or the insect's receptors recognize the shape of the signal molecules, and in particular the distribution of electronic charge on their surfaces. The simplest analogy is of a key fitting precisely into a lock, but in this case the key must have not only the right shape but also the right distribution of electrical charges on its surface. In this respect, the magnetic card locks that are common at hotels are better representations of molecular recognition in that they not only detect the key but also flash a green light to show you when they have detected the right key (and are thus true sensors since they both detect the key and give a signal).

In addition to chemical sensing, the biological world relies on sensors for other properties. Many flowers and leaves are attracted to the sun, which is their source of energy. Particular molecular sensors in the structure of the leaf or flower respond to the presence of the sun. These sensors signal the molecular motor structure of the leaf or flower to move in a particular direction, to face the sun and gain more energy from it. We'll talk about this much more in Chapter 9. Animals have ears to sense sounds, and fish have lateral lines to sense sound and pressure variations. All of these are sensing mechanisms and all of them are crucial to life.

Sensors in the synthetic nanoworld will prove to be just as critical and are often built on the same premises as their natural counterparts rather than their current artificial big brothers, macroscale sensors. Those macroscale artificial sensors usually work via physical properties of bulk materials or via complex mechanical or electronic apparatus. For example, thermometers work by measuring the thermal expansion of liquid mercury and accelerometers use microelectromechanical systems to measure the acceleration or deceleration of a car. Neither MEMS nor thermal expansion is an easy process to translate to the nanoscale. Instead, nanoscale sensors will often either mimic those life processes that have already developed in the nano-world, or use key quantum mechanical or size-dependent physical properties that exist only there. This means not only that nanoscale sensors will be the best and most accurate sensors possible, but also that they will be able to sense things that simply could not be detected by macroscale devices.

Synthetic sensors can be classified according to what they sense—we will discuss sensors for electromagnetic radiation, for small and medium-sized molecules (like the squares we discussed in Chapter 5), and for biological entities.

ELECTROMAGNETIC SENSORS

For our purposes, the term "electromagnetic" refers to any form of energy that is propagated as a wave. Starting with the lowest energy and going to the highest, some examples are radio waves, infrared light, visible light from the red to the violet, ultraviolet light, and x-rays. Sound is essentially a propagating pressure wave, and therefore slightly different from electromagnetic radiation, but it is sensed in a very similar fashion.

The simplest electromagnetic sensors respond to a physical condition, like photoelectric cells that are used to turn lights on when the sun goes down. These work by measuring the intensity of light coming from the sun; for example, when a bright light drops below a certain predetermined brightness level, a signal is given for the electricity to be turned on.

To develop a nanoscale photosensor, it is possible to ride on the back of research in solar power generation. In Chapter 5, we discussed the development of the photoelectrochemical cells, such as those developed by Michael Graetzel for capturing sunlight. These cells use molecular dyes that are excited by capturing sunlight. The excited molecules then transfer an electron into a nanoscale quantum dot of a semiconductor like titanium dioxide. Using one of these photoelectric devices as a sensor is straightforward. It is simply necessary to measure that the electron has been transferred, and this is relatively easy to do because the transferred electron moves through an external circuit to lower its energy by recombining with the positive charge left behind on the dye. In other words, if a photocell is producing electricity, we know that it has been exposed to light.

A very much older, and in some senses even more elegant, nanoscale photosensor is behind the science and art of photography. In traditional silver photography, the photons (light energy) cause a chemical reaction among the silver ions that are held in the emulsion on the film surface. The silver ions come together to form nanoscale silver clusters (the simplest is actually just four atoms) that grow large enough to scatter and capture the light, thereby appearing black on the surface. Again, the change in properties with size that is so essential to nanotechnology is at work here.

Making x-ray, ultraviolet, or infrared film requires very similar processes. (Often those films are made by the same companies that make photographic film.) The requirement is simply that the photoactive agent, which is still often silver, must interact with light of the appropriate wavelength. For x-rays, the wavelengths are very much shorter; for infrared, they are very much longer. To tune a Graetzel cell-based photosensor to respond to different colors or types of light, it is only necessary to find an appropriate dye molecule.

The sensors in film are molecular/atomic, and the sensing process consists of an irreversible change in the clustering of the silver atoms. Microphones sense sound or pressure in a very different way. They consist of diaphragms that are set to vibrating when exposed to a pressure or sound wave. This is very much the same principle as a drumhead—when an external source of pressure hits the drumhead, it starts to vibrate. In fact, hair cells in our ears work in this same way: the membrane is set into vibration by the external pressure wave that corresponds to sound, and then a very sophisticated set of chemical signals is activated by the vibrations of the membrane. The ear is a very complex multiscale electromagnetic sensor that is based largely on molecular signals. The energy is changed and propagated (the technical term is *transduced*) from the vibrations in the membrane into electrochemical signals that travel to the brain.

Because most electromagnetic sensors are already designed to deal with nanoscale or near-nanoscale waves, shrinking them tends to be less complicated than it is for other kinds of sensors. It is an interesting part of nanotechnology, but not quite as groundbreaking as nanoscale artificial biosensors, an entirely new field.

BIOSENSORS

Biosensors are not just natural sensors that are part of life; they are sensors for biological entities including proteins, drugs, and even specific viruses. Nature does have a variety of schemes for approaching the detection of these entities. One common method is the one behind allergic response. When a body is first exposed to an *allergen* (a benign substance that it mistakes for a hostile invader), it is sensitized, which means that it creates antibodies that will recognize that allergen if it ever appears again. The antibodies use molecular recognition to spot the allergenic proteins and release *histamine*, the substance that causes your body to react by sneezing, itching, or nausea. This ability to sense larger structures, such as proteins or nucleic acids, can be remarkably important.

Glucose detection is a classic problem in biosensing. Diabetics cannot control their insulin levels; therefore, their levels of blood glucose fluctuate tremendously. If those levels get either too high or too low, their conditions can be life threatening. Currently, most Type I diabetics must actually draw blood on a daily basis, or even more often, to test for blood glucose levels. Sensing glucose molecules can be done in many ways, using optical, conduction, or molecular recognition methods. None of these have yet been shown to be compatible with an implantable simple device that could automatically, continuously sense the glucose levels in the blood. This is one of the major challenges for chemical sensing, and nanoscale structures may advance it very substantially.

DNA sensing is potentially an enormous area in which nanoscience can improve medicine. When we discussed DNA computing in Chapter 5, we talked about hybridization, the ability of DNA to bind to a complementary strand and not to bind to anything else. For instance, if we wish to sense the structure with the sequence CGCGTTC, we could do so by using a strand GCGCAAG. This means that a single strand of, say, six bases can contain 4,096 different combinations (each base can have one of four values A, C, G, or T, so a six-base sequence can have 4 x 4 x 4 x 4 x 4 x 4 possible values). Consequently, if a particular biological target such as botulism or strep or scarlet fever has a known DNA sequence, it is possible to target a short section of that DNA sequence—say a section of 10 to 15 bases—that can be uniquely sensed, without any errors, by an appropriate single-strand complementary structure. This is sometimes called a DNA fingerprint for a disease because it is virtually impossible to mistake an analyte using even a moderately long sequence. With a 15 base strand, there is only a one in one billion chance of error per strand tested.

The most striking application of DNA sensing will probably come in the generalization of the lab-on-a-chip concept. By using the powerful analytic capabilities of these dense microlaboratories, it will be possible to include several screening sensors on a chip for instant recognition of viral or bacterial DNA associated with several different diseases found in the body. Such chips could also be used to sense the presence of toxic species, either natural or artificial, in water supplies. Finally, since we now know the entire human genome, biochips could be used to sense either particular DNA signatures or particular protein signatures known to be defects that can result in disease. This would allow at-risk individuals to receive more frequent tests and attention. Multiple-sensing or sequencing of DNA is a major target in the biomedical community because it will permit tremendously advanced diagnostics. DNA sensors are clearly going to be the optimal (and maybe the only) way to respond to this challenge.

It is also possible to create sensors that take advantage of DNA recognition. The simplest sensors work by introducing a strand of DNA complementary to the analyte into a solution to be tested. If the analyte is present, it will hybridize with the test DNA and form a double strand.

Hybridization confirms that the analyte is present; finding out whether hybridization has occurred isn't trivial. We can't see the double strands without very sophisticated instruments; consequently, the determination is usually made by mass. Obviously double strands have a greater mass than single strands, though it may not be by much if the test sequence is short since each base pair weighs only as much as a molecule or roughly 1/1,000,000,000,000,000,000,000 of a gram. This is much too small to measure easily in any direct way, so it is necessary to amplify the response before it can be measured. One of the great challenges of DNA sensing is therefore to amplify the effects of hybridization so that they can be easily measured.

One way to provide this amplification is to change the optical properties of gold or silver nanodots that are attached (technically "functionalized") to the DNA. Chad Mirkin, Robert Letsinger, and their groups at Northwestern pioneered the combination of quantum optical effects (remember the changes in the color of gold upon changing the size of the gold clusters) and molecular recognition (complementary DNA binding). Their scheme and some actual results are shown in Figure 7.1.

By exposing the single strands of DNA that are attached to the gold nanodots, the sensor recognizes the target strands of DNA, which causes the gold nanospheres to come closer together and, as in those recurring stained glass windows, change color. Because of the color change, these are called *colorimetric sensors* and can be read by simply looking at them. (George Schatz's group at Northwestern has worked out the theoretical basis.) This approach is typical of the kinds of sensing methods that are used for looking for tiny fragments of DNA.

In addition to DNA, colorimetric nanosensors can use the change in color of metallic nanodots to detect other molecules. Richard Van Duyne's group at Northwestern has used nanosphere lithography to prepare tiny gold dots on a surface, as we discussed in Chapter 4. A molecular nanostructure containing a biological binding site (something like an antibody) is attached to the gold nanoparticles. The binding site is designed to recognize (chemically bind to) a particular protein analyte just as antibodies bind to biological invaders in your body. When that analyte appears in solution, it binds to the recognition site, which changes the chemical and physical environment of the gold dot, whose color is then slightly changed. This change can be measured, and the sensitivity is enormous. Van Duyne has shown that his gold nanodots can actually measure a single molecule of particular analytes.

The challenge for creating totally generalized nanosensors of this type is that, in order to be useful, sensors must not create false positive results. In the case of DNA, it is almost impossible for hybridization to occur and for the sensor to trigger if the target molecule is not present. But if you wanted to construct an explosives detector, the problem is much more complex. Nitrates, which are the molecular groups common to most explosives, are common in a variety of other common household items including hot dogs and fertilizer and can even be found within the human body. If you detected them to an accuracy of a single molecule, every single person tested would seem to be carrying a bomb. A great deal of research is being done to circumvent this problem for explosives and other common analytes.

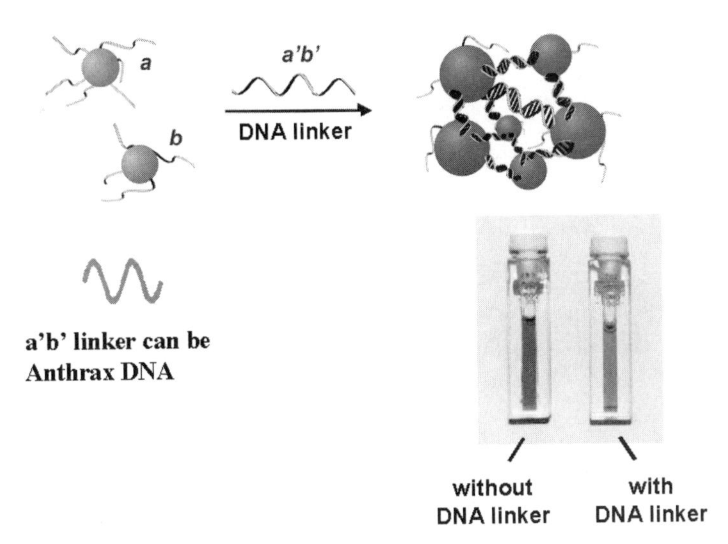

Figure 7.1
The upper schematic shows how the nanodots in a colorimetric sensor are brought together upon binding to the DNA target (in this case anthrax). The clustered dots have a different color than the unclustered ones as is shown in the photograph below them. *Courtesy of the Mirkin Group, Northwestern University.*

ELECTRONIC NOSES

We looked at how biological noses work using molecular recognition to send a neural signal to the brain. In the artificial nose, the most common replacement for the nasal membrane is an electrically conducting polymer. When the polymer is exposed to any given molecule in the vapor phase, its conductivity will change a little bit. In the electronic nose, a random polymer, or mix of polymers, is spread between electrodes. When the molecules to be smelled land on the polymer(s), the conductivity properties in particular regions will change in a particular way that is specific to any given analyte. The nature of the detection is interesting, since it is based on so-called neural nets. The idea is that each sensor will give a particular voltage and current signal, and these are then compared with the list of standard signals that the nose has already smelled. In this sense, it is necessary to "train" the nose, just as we require training. When a child asks what's the funny smell and mom says it's a skunk, the child has learned. By measuring the response of the electronic nose to a series of standard molecular inputs, we can determine the electrical signal caused by particular analytes. Then, when an unknown vapor causes the electrical system to respond in the same way, we can deduce that the analyte is present. Since the electronic properties of the polymer molecules are key to electronic noses, there is some crossover here between sensors and one of our next topics, molecular electronics.

There are already several commercial companies marketing electronic noses (with names like Cyrano) for applications as different as toxic gas detection, disease analysis, air quality monitoring, and food inspection and standardization.

Chad Mirkin, nanoscientist and entrepreneur, envisions a world where doctors will be able to diagnose diseases and conditions in the examination room instantly and accurately not only by analyzing a patient's symptoms but also by actually sensing pathogens in a patient's body or blood. Together with renowned researcher Robert Letsinger, Mirkin founded a startup company Nanosphere, Inc., which has begun commercializing their work on biosensors. Using the same DNA-matching biosensor techniques that we discussed, they have already developed one of the world's quickest and most accurate tests for anthrax and are developing a suite of tests for other diseases including AIDS. If Nanosphere succeeds in developing its planned products (prototypes already exist), medical technicians will be able to administer all common tests right in the office or at the bedside. Additionally, military and law enforcement officials will be able to check for contamination of letters, buildings, and battlefields more easily. By integrating biosensors into a lab-on-a-chip, many tests can be run in parallel and perform a single-step screening for a variety of diseases instead of individual tests for tuberculosis, hepatitis, measles, mumps, and others as is done now. With luck, they will soon create a rabies and tetanus test to eliminate the need to treat these diseases with painful injections just as a precaution. By using these technologies, Nanosphere and other nanotechnology-based sensor companies may be able to make disease screening cheap and easy enough that it could be comprehensive even in low-income countries. This is one part of the dream of nanotechnology that will completely change the way things are done and may be reality as soon as the next two to three years.

8 Biomedical Applications

In this chapter...

Paracelsus, the doctor who helped break through stagnant medieval thought and revolutionize medicine in the 16th century, is credited with being the first to use minerals and inorganic compounds to treat human diseases. Before Paracelsus, most medieval doctors believed that it was necessary to treat organic problems with organic drugs, which gave them a limited arsenal of plant and animal therapies to choose from. After him, medicine thrived and became much more scientific and effective.

Nanoscience promises to take Paracelsus' idea to the next level by incorporating not just inorganic therapies, but inorganic concepts. Biomedicine already had a second renaissance with the discovery of DNA and our better understanding of how diseases and harmful conditions work, but it has not yet been able to use all that knowledge effectively to create treatments for many of humanity's most serious problems. Nanoscience will help capitalize on that knowledge, taking advantage of developments in the physical sciences and in nanoelectronics. The nanoscale is the natural scale of all fundamental life processes, and it is the scale at which diseases will need to be met and conquered.

Developments in nanoscale biomedicine have the potential to create implants to monitor blood chemistry and release drugs on demand, which could be a boon for diabetics and cancer patients and could result in better forms of birth control. Artificial bone, cartilage, and skin that will not be rejected by the human immune system are being developed, as are more radical forms of human repair. Advances in cancer treatment are almost certain and may take the form of improved, directed chemotherapy and radiotherapy or even an outright cure for some forms of cancer. The fusion of interdisciplinary knowledge coming together at the nanoscale will be one of the great benefits nanoscientists will introduce into our lives.

DRUGS . . . .

Drugs come in many categories. Some are simple continuous structures such as creams or lotions that change the properties of the skin or control its exposure to external elements such as sunlight. Some are large, macromolecular biological structures that specifically interact with other large macromolecular structures in the body—examples include vaccines, which are basically modified viruses. One of the largest and most important categories comprises individual molecules that work by interacting specifically with DNA or proteins within the body. This class of drug molecules includes aspirin, the anticancer drug Cis-Platin, and much more complex molecules like beta blockers or antiinflammatories such as cortisone.

Since most of these molecular drugs are nanosize, drug development is clearly a nanoscale activity. It becomes even more so when the drugs are designed specifically to interact with known biological targets. This process of nanoscale drug design is rooted in mechanistic ideas about the biological target of the drug. For example, because depression is often caused by too low or too high a concentration of *neurotransmitter molecules* (the molecular couriers that carry messages between synapses in the brain), intelligent nanoscale development of antidepressants is focused on increasing this concentration by blocking or decreasing the destruction of these molecules by modifying their binding properties. Restoring the concentration to normal levels is an effective way to treat the symptoms of some forms of depression. The same mechanistic approach to drug therapy has had a number of successes, such as in the development of drugs now used in AIDS therapy.

Nanoscale methods, both computational and experimental, are helping to make the drug discovery process less one of happenstance or empirical evidence and more one of design. In Paracelsus' day and even now, many drugs are used just because they seem to work even when the mechanism is not well understood. This is part of the reason why side effects often aren't found until clinical testing or even until after a drug is on the market. It's also part of the reason for the long and complex drug approval process of the FDA. Since many proposed nanodrugs will work by well-understood and very specific mechanisms, one of the major impacts of nanoscience and nanotechnology will be in facilitating development of entirely new drugs with fewer side effects and more beneficial behavior.

DRUG DELIVERY . . . .

People are very large compared to molecules, and it is important for therapeutic effectiveness that the drug molecules find the places in the body where they will be effective—antidepressants should be in the brain; anti-inflammatories at sites of stress; and anticancer drugs at the tumor sites. *Bioavailability* refers to the presence of drug molecules where they are needed in the body and where they will do the most good. The issue of drug delivery centers on maximizing bioavailability both over a period of time and at specific places in the body. In fact, over $65 billion of current pharmaceuticals suffer from poor bioavailability.

Increasing bioavailability is seldom as simple as increasing the amount of a drug that is used. In chemotherapy, for example, the drugs used are actually somewhat toxic, and increasing the amount used could adversely affect or even kill a patient. On the other hand, if the drug could be delivered directly to the site of a tumor before the tumor *metastasizes* (the process where a tumor spreads to adjacent organs or the blood) and without interacting with the rest of the body, chemotherapy could become more effective and much less unpleasant.

Nanotechnology and nanoscience are very useful in developing entirely new schemes for increasing bioavailability and improving drug delivery. For example, molecules can be encapsulated within nanoscale cavities inside polymers. The polymer can then be swallowed as part of a tablet, and as the polymeric structure opens within the body, the enclosed drugs can be released. This is an effective method for creating time-released drugs so that a pill taken once a day or once a week can continue to deliver the drug smoothly over an extended period of time. Another even simpler scheme, which has the effect of maximizing overall bioavailability for a short time, is to grind solid drugs into fine powders, sometimes so fine that the particle sizes are in the nanoscale. This is done to increase surface area and reactivity—the same reasons that catalyst particle sizes (like those in Chapter 6) are reduced—and to increase solubility in the body.

Much more complex drug delivery schemes have also been developed, such as the ability to get drugs through cell walls and into cells. Efficient drug delivery is important because many diseases ranging from sickle cell anemia to Wilson's disease depend upon processes within the cell and can only be interfered with by drugs delivered into the cell. Many drug molecules cannot pass through the membrane that surrounds the cell, essentially because of difficulty associated with putting polar molecules into the nonpolar membrane. One way to get around this problem is to encapsulate the polar drug in a nonpolar coating that will easily pass through the cell membrane. For example, small molecules of DNA that combine with alien pathogenic DNAs within the cell can be used as a drug. To make these artificial DNA drugs highly available within the cell, however, you must pass them through the membrane. One way of doing this is to coat the DNA with cholesterol. Cholesterol is a fatty hydrophobic molecule that easily passes through the oily cell membrane. By enclosing the DNA drug within a blanket of cholesterol, it can be delivered into the cell where it can be most effective, showing that sometimes cholesterol is good for you. *Liposome structures* based on balls of fatty molecules enclosing the drug work similarly. Liposomes have been used in cancer treatment, delivering soluble proteins (cytokines) such as interferon to cancer cells.

Magnetic nanoparticles of the sort described in Chapter 5 for computer memory can also be used for drug delivery. Here, the nanomagnet is bound by a molecular recognition method to the drug to be delivered. Then magnetic fields external to the body can manipulate the position of the nanodot, and therefore control local bioavailability of the drug. Effectively a doctor can drag the drug through the body in the same way that you drag iron filings across a table with a hand magnet.

One interesting combination of smart materials and drug delivery involves triggered response. This consists of placing drug molecules within the body in an inactive form that "wakes up" on encountering a particular signal. A simple example would be an antacid enclosed in a coating of polymer that dissolves only in highly acidic conditions; the antacid would then be released only when the outer polymer coat encounters a highly acidic spot in the digestive tract.

Artificial bone materials like those discussed in Chapter 5 are another example. The molecules that come together to form the artificial bone cylinder can be placed either inside or outside the body and can be programmed by design to come together to form the rigid cylinder only when exposed to a trip signal, which might be something as simple as exposure to liquid water or a cut or an impact. They might operate in much the same way as platelets in the bloodstream.

Molecular design and molecular nanobiology are resulting in many new smart drugs. Two particularly intriguing examples are the so-called suicide inhibitors and DNA molecule therapies. *Suicide inhibitors* are not actually intended to discourage suicidal behavior among people as the name implies, though they can be used to treat depression. Instead, they are designed to block the action of certain enzymes by causing the enzymes, in effect, to commit suicide. They start their journey as benign molecules structured so that the enzyme they are supposed to destroy recognizes them and tries to do its normal job and modify them. But these molecules are slightly different than most molecules that the enzyme modifies in that the modification results in a new molecule with an exposed, highly energetic end that is just looking for something to bind to. What it binds to is the enzyme itself. The resulting bond is so strong that it is effectively irreversible. The new combined structure (enzyme plus drug molecule) does not function like the enzyme alone; consequently, the enzyme, by performing its normal function, has committed suicide. Pharmaceutical companies have used some of these drugs, such as those developed by Richard Silverman's group at Northwestern, to deal with conditions such as epilepsy and depression, both of which have an important enzyme action component. Suicide inhibitors limit enzymatic activity and, thereby, can be efficacious in eliminating the symptoms of disease.

The other example is *DNA molecule therapy* which is a form of gene therapy, which takes advantage of DNA's unique properties of self-binding. When we looked at DNA sensors, in Chapter 7 we discussed how it is possible to detect a given biological entity by creating a complement for its DNA fingerprint and seeing if the complement successfully finds the fingerprint and binds. Under some conditions, it is possible that this binding can be made irreversible, which means that the disease-causing DNA can be prevented from ever replicating again and thus be removed as a threat. Because the drug DNA won't bind to anything but its target, it is also completely safe for the person using it. This approach could potentially be used for creating antiviruses, although the fact that viruses mutate will remain a challenge. All of the many methods of approaching DNA molecule therapy rely on the chemical synthetic techniques for making the complementary DNA strands that we discussed in Chapter 4. DNA molecule engineering is one of the most active areas of bionanoscience, both for its biological applications and because the complement binding in DNA provides a remarkable and uniquely effective way of assembling nanostructures that can be used for many purposes beyond their medicinal value.

PHOTODYNAMIC THERAPY . . . .

Nearly every child has held up a flashlight to her hand and seen how the light passes through her hand and looks red when it comes out. This red color isn't only because of red blood. Instead, like the blue color of the sky, it arises because the amount of light that is scattered by an object depends on the wavelength of the light. Light with long wavelengths can actually pass through biological tissue without excessive scattering and, therefore, can be used to affect processes within the body.

In *photodynamic therapy*, a particle is placed within the body and is illuminated with light from the outside—the light could come from a laser or from a light bulb. The light is absorbed by the particle, after which several things might happen. If the particle is simply a metal nanodot, the energy from the light will heat the dot and, therefore, will heat any tissue within its neighborhood. With some particular molecular dots, light can also be used to produce highly energetic oxygen molecules. Those oxygen molecules are very reactive and will chemically react with, and therefore destroy, most organic molecules that are next to them, including such nasties as tumors.

These therapeutic ideas are referred to as photodynamic therapy because they not only are promoted by light (photons) but also depend on the excited state dynamics of the molecules or dots involved. Photodynamic therapy is attractive for many reasons, one of which is that, unlike traditional chemotherapy, it is directed. The chemically reactive excited oxygen, or even the heat from excited quantum dots, is released only where the particles are and where the light is shined. This means that, unlike traditional chemotherapy, photodynamic therapy does not leave a "toxic trail" of highly aggressive and reactive molecules throughout the body.

The particles involved in photodynamic therapy can be as simple as medium-sized molecules that can excite oxygen when illuminated. They can also be as complicated as a multicomponent quantum structure, part of which "recognizes" (or binds to) a target species such as a tumor, while the remainder absorbs the radiation and either heats or provides excited oxygens. There can be other components to provide bioavailability.

Photodynamic therapy is based on nanostructures ranging from simple molecules through molecule/nanoparticle/biological recognition agent composite structures. Clearly the design and optimization of such structures is a matter of medical nanotechnology and holds promise as a noninvasive approach for dealing with many growths, tumors, and diseases.

MOLECULAR MOTORS

Within large biological structures such as individual cells, different components have to move. Sometimes molecules and ions and even larger biological structures move by simply diffusing. Some examples include the motion of neurotransmitters in the brain and of ions through ion channels in the outer membrane of the cell. As molecular species get larger, their diffusion becomes less efficient, and so nature has developed a series of special mechanisms to move them within the cell. One of the most fascinating of these is the molecular motor.

Molecular motors were discovered during the study of one of the energy-generating functions of the body called sodium/potassium ATPase. This is a complex enzyme that is responsible for producing and transducing energy stored in the molecule ATP, which is the common energy currency of the body powering everything including muscle movement. Sodium/potassium ATPase actually functions as a rotary motor: the central unit of the nanostructure rotates around a pivot, and the outside part of the nanocluster reacts differently with chemical groups around the periphery.

This rotary motion is one of several molecular motor mechanisms that are now understood to play important roles in functional biology of the cell. Molecular motors also allow us to manage the availability of different components of the cell, as they move about within

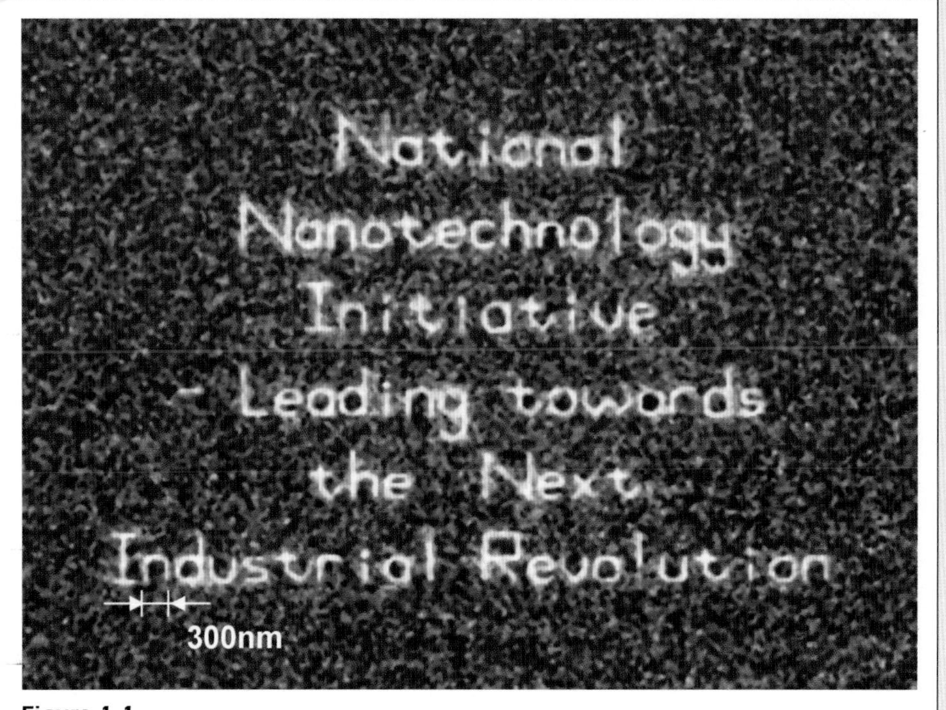

Figure 1.1

The Next Industrial Revolution, an image of a nanostructure.

Courtesy of the Mirkin Group, Northwestern University.

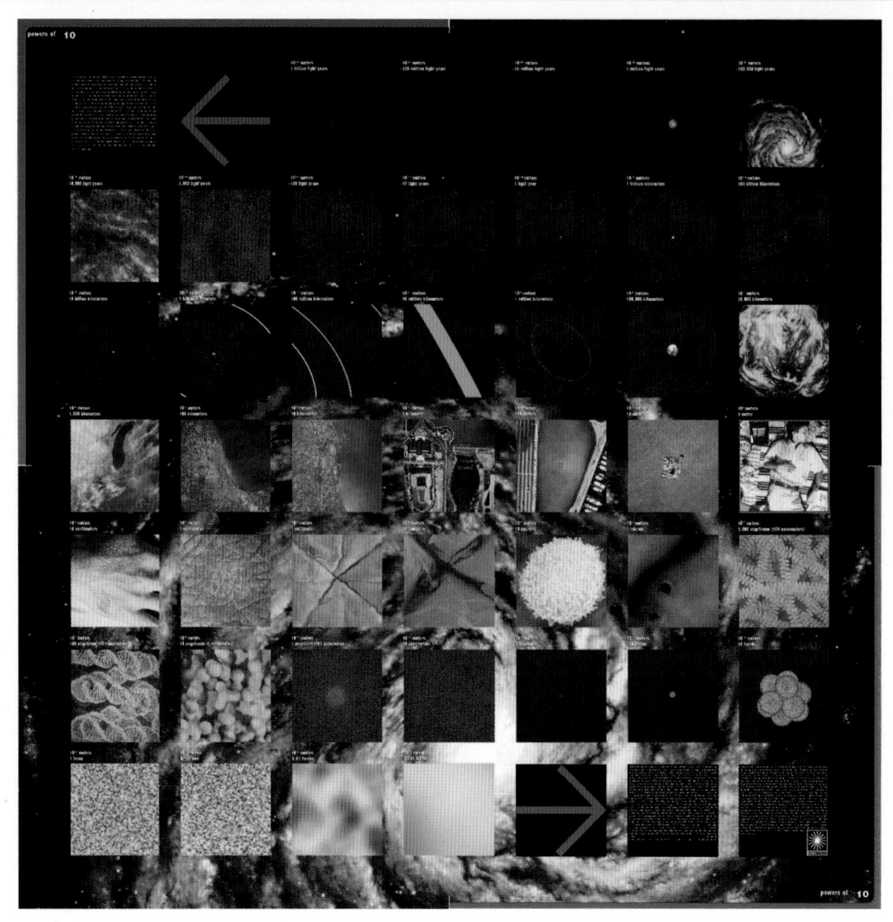

Figure 1.2
This image shows the size of the nanoscale relative to things we are familiar with. Each image is magnified 10 times from the image before it. As you can see, the size difference between a nanometer and a person is roughly the same as the size difference between a person and the orbit of the moon.
©2001 Lucia Eames/Eames Office (www.eamesoffice.com)

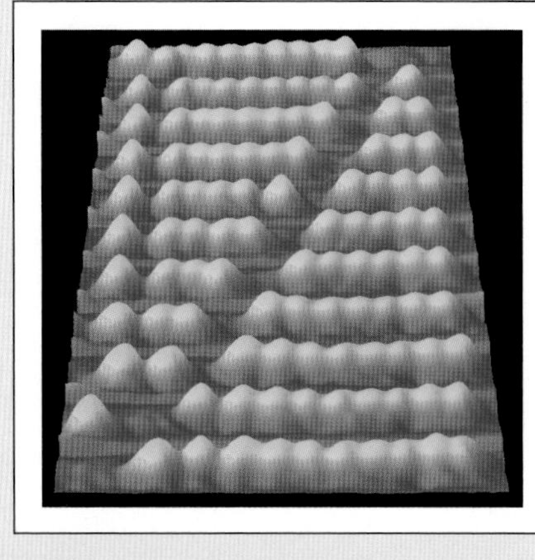

Figure 1.3
The nanoscale abacus. The individual bumps are molecules of carbon-60, which are about 1 nanometer wide.
Courtesy of J. Gimzewski, UCLA.

Figure 2.1
Early nanotechnologist.
Courtesy of Getty Images.

Figure 2.2
Modern nanotechnologist.
Courtesy of Getty Images.

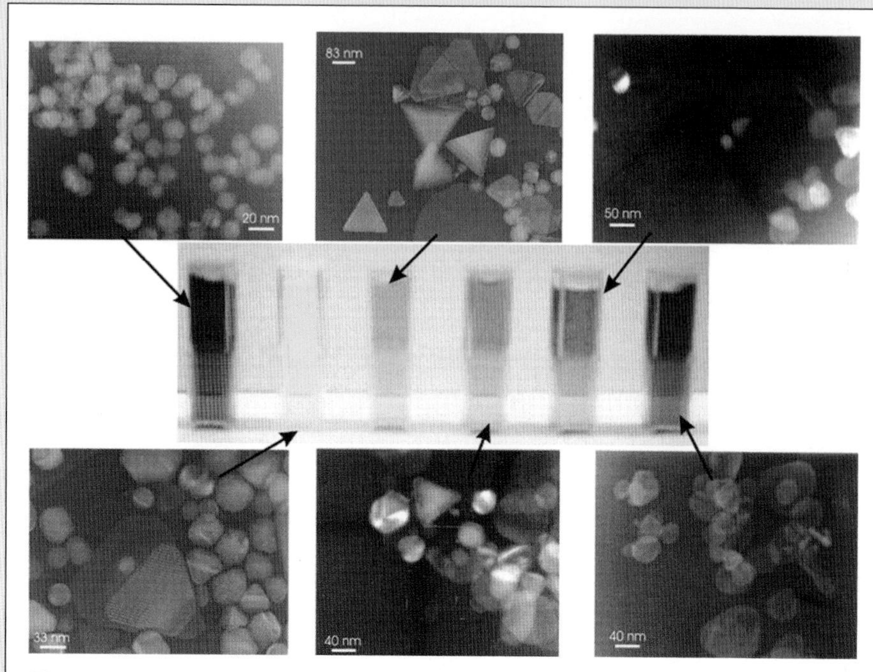

Figure 2.3

Nanocrystals in suspension. Each jar contains either silver or gold, and the color difference is caused by particle sizes and shapes, as shown in the structures above and below.

Courtesy of the Van Duyne Group, Northwestern University.

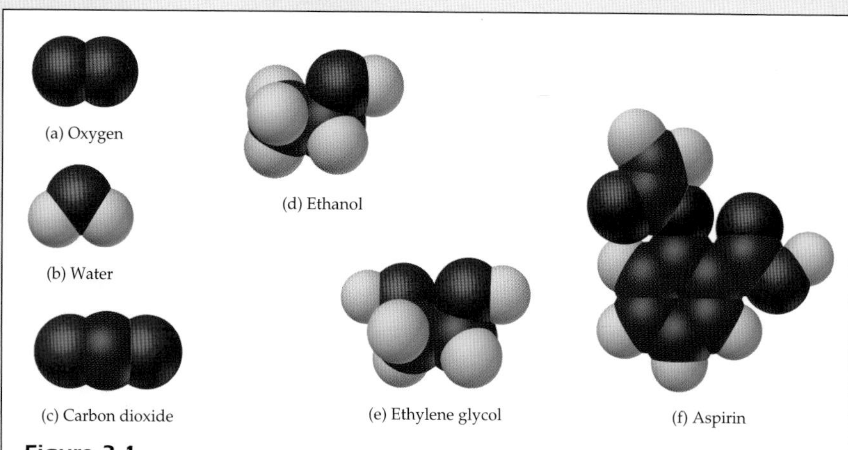

Figure 3.1

Models of some common small moledules. The white, gray, and red spheres represent hydrogen, carbon, and oxygen, respectively.

From Chemistry: The Central Science, *9/e, by Brown/LeMay/Bursten, © Pearson Education, Inc. Reprinted by permission of Pearson Education, Inc., Upper Saddle River, NJ.*

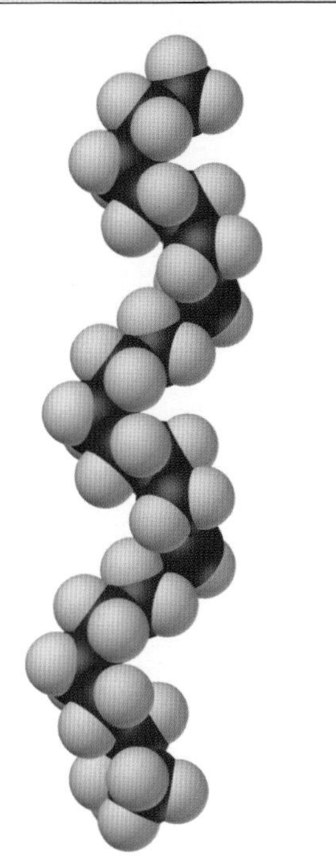

Figure 3.2
A molecular model of a segment of the polyethylene chain. This segment contains 28 carbon atoms (dark), but in commercial polyethylene there are more than a thousand carbon atoms per strand.

From Chemistry: The Central Science, *9/e, by Brown/LeMay/Bursten, © Pearson Education, Inc. Reprinted by permission of Pearson Education, Inc., Upper Saddle River, NJ.*

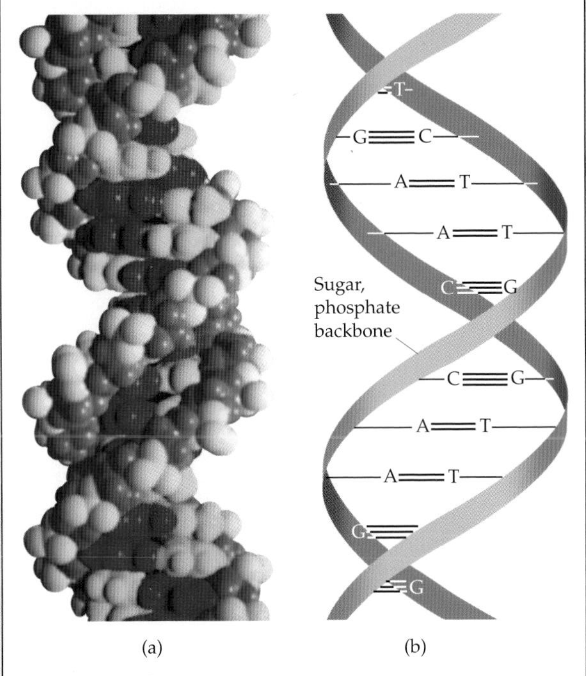

Figure 3.3
(a) Computer-generated model of the DNA double helix. (b) Schematic showing the actual base pairs linked to each other.

From Chemistry: The Central Science, *9/e, by Brown/LeMay/Bursten, © Pearson Education, Inc. Reprinted by permission of Pearson Education, Inc., Upper Saddle River, NJ.*

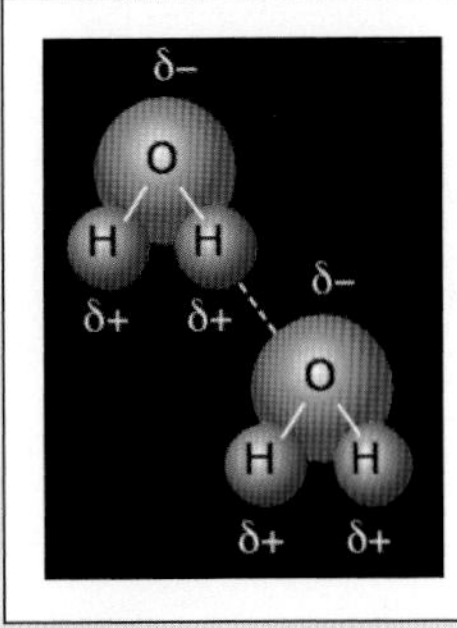

Figure 3.4
Molecular binding of two water molecules. The symbols δ+ and δ- denote positive and negative charges, respectively.

Courtesy of the Advanced Light Source, Lawrence Berkeley National Laboratory.

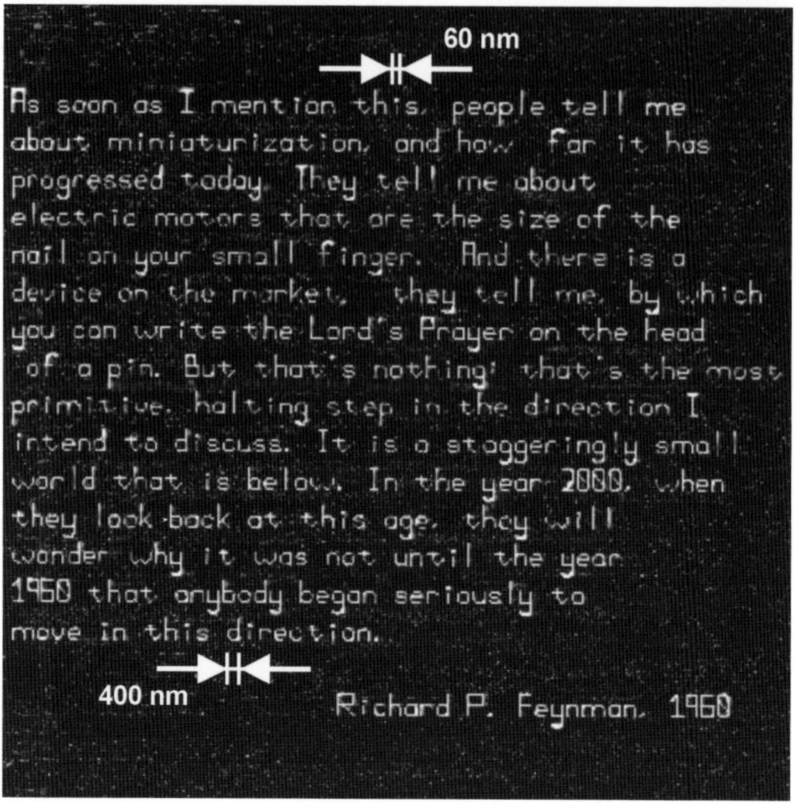

60 nm
As soon as I mention this, people tell me
about miniaturization, and how far it has
progressed today. They tell me about
electric motors that are the size of the
nail on your small finger. And there is a
device on the market, they tell me, by which
you can write the Lord's Prayer on the head
of a pin. But that's nothing; that's the most
primitive, halting step in the direction I
intend to discuss. It is a staggeringly small
world that is below. In the year 2000, when
they look back at this age, they will
wonder why it was not until the year
1960 that anybody began seriously to
move in this direction.
400 nm
Richard P. Feynman, 1960

Figure 4.1

The founding speech of nanotechnology—written at the nanoscale.

Courtesy of the Mirkin Group, Northwestern University.

100 microns
Mag =300 X Tip A1

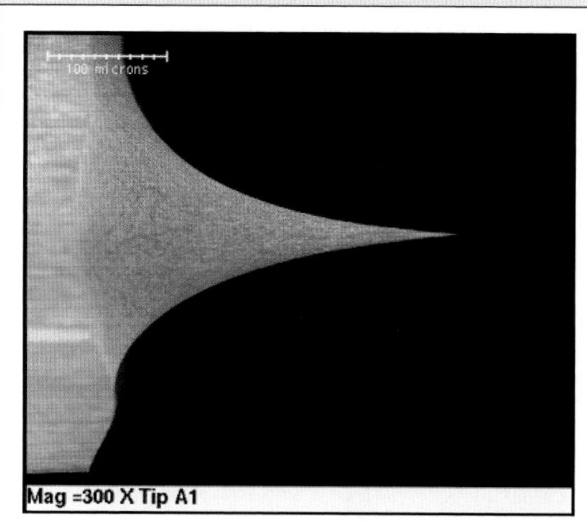

Figure 4.2

An STM tip made of tungsten.

Courtesy of the Hersam Group, Northwestern University.

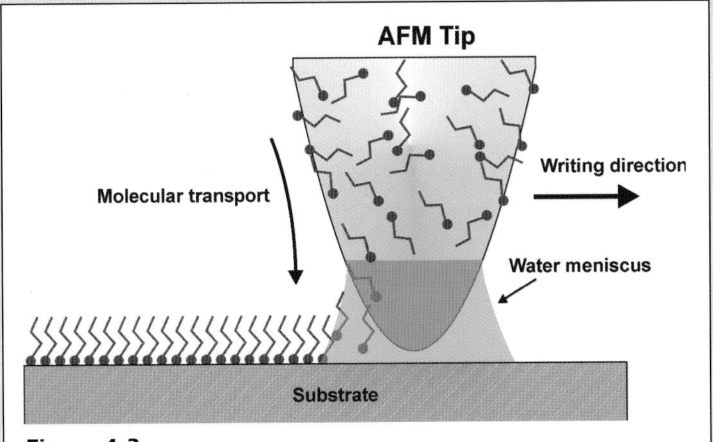

Figure 4.3
Schematic of the dip pen lithography process—the wiggly lines are molecular "ink."
Courtesy of the Mirkin Group, Northwestern University.

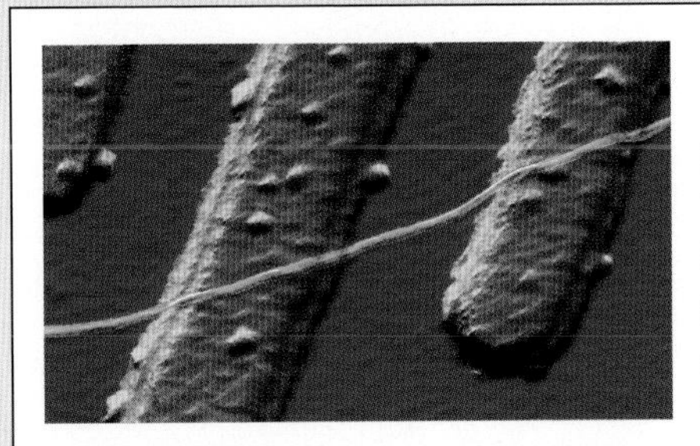

Figure 4.4
Two electrodes made using E-beam lithography. The blue structure is a carbon nanotube.
Courtesy of the Dekker Group, Delft Institute of Technology.

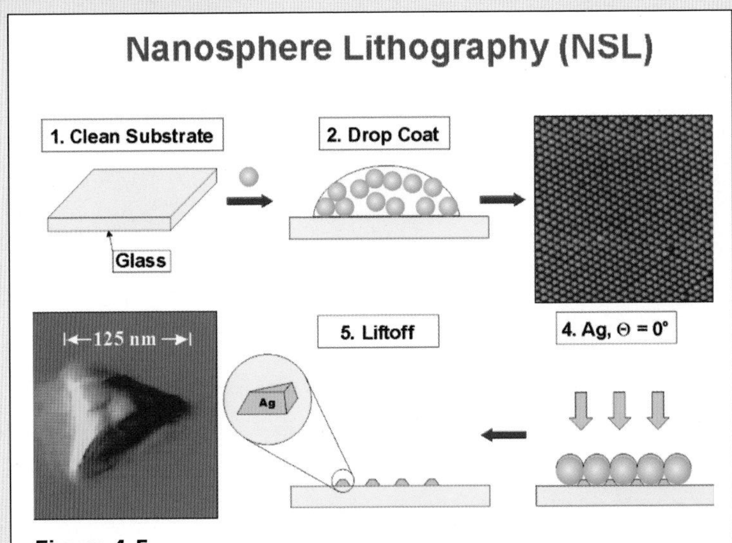

Figure 4.5
Schematic of the nanosphere liftoff lithography process.
Courtesy of the Van Duyne Group, Northwestern University.

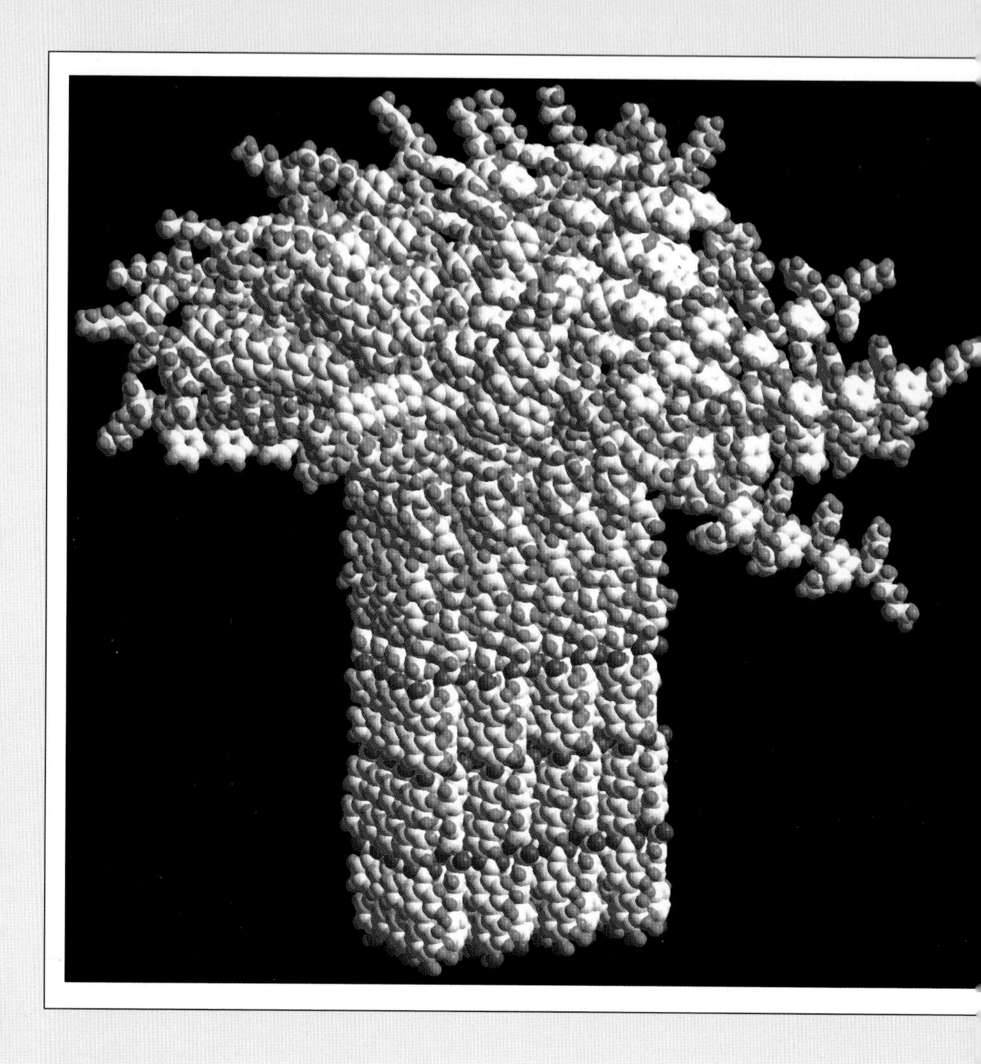

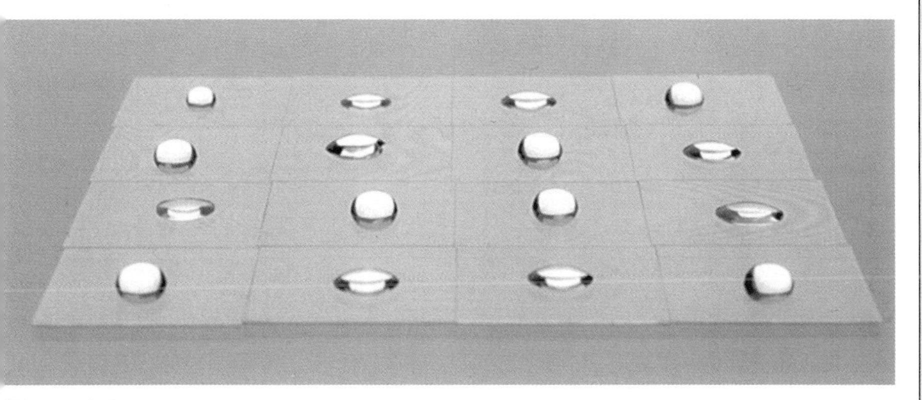

Figure 4.6
Molecular model (left) of a self-assembled "mushroom" (more correctly a rodcoil polymer). The photograph (right) shows control of surface wetting by a layer of these mushrooms.
Courtesy of the Stupp Group, Northwestern University.

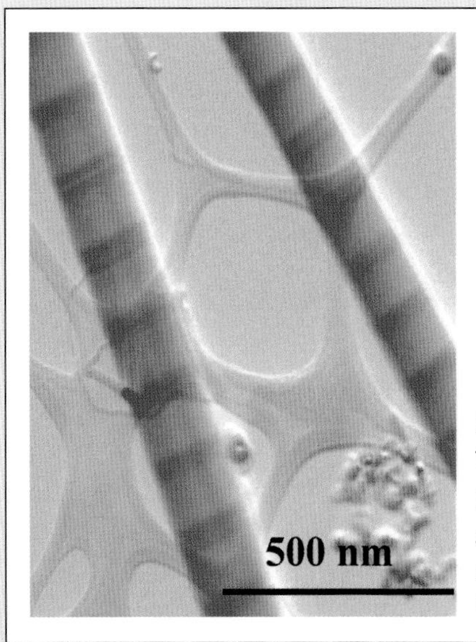

Figure 4.7
Two parallel nanowires. The light color is silicon, and the darker color is silicon/germanium.
Courtesy of Yang Group, University of California at Berkeley.

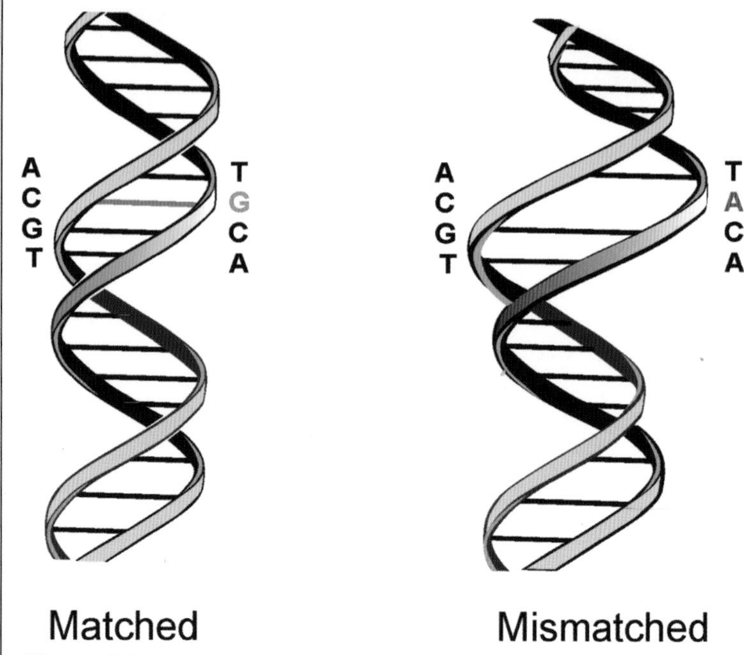

Figure 4.8

Schematic of the DNA hybridization process. The "matched" side shows how a DNA strand correctly binds to its complement and the "mismatched" side shows how errors can prevent binding.

Courtesy of the Mirkin Group, Northwestern University.

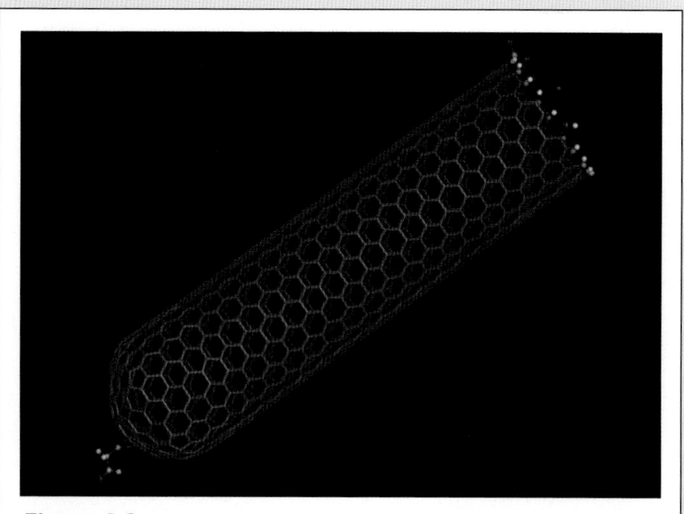

Figure 4.9

A singled-walled carbon nanotube.

Courtesy of the Smalley Group, Rice University.

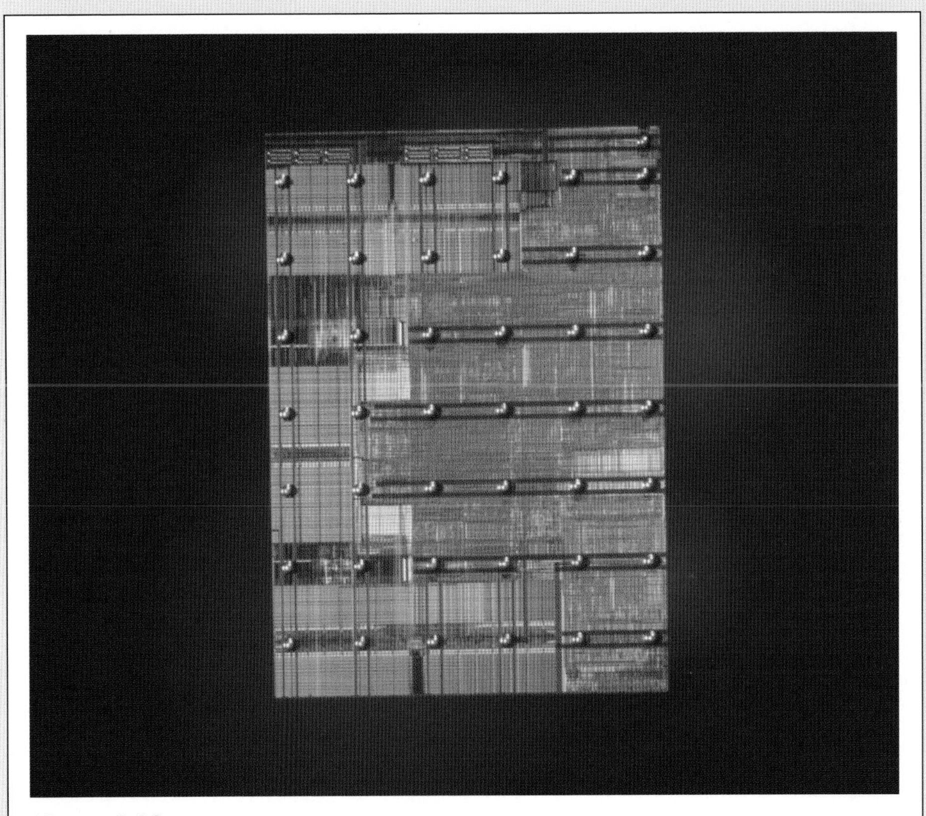

Figure 4.10
A current CMOS chip surface.
Courtesy of Tom Way/IBM Corporation.

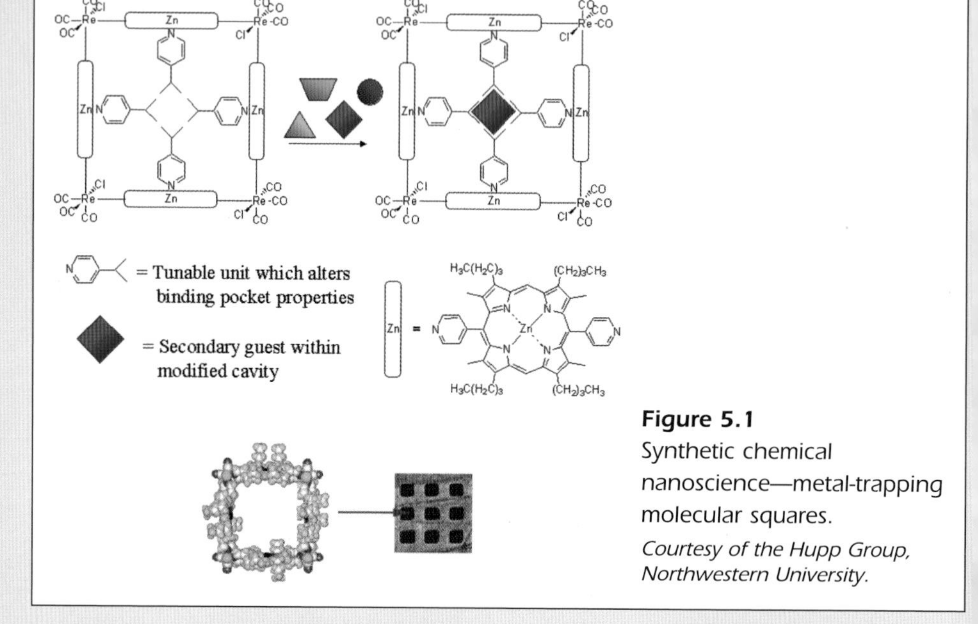

Figure 5.1
Synthetic chemical nanoscience—metal-trapping molecular squares.
Courtesy of the Hupp Group, Northwestern University.

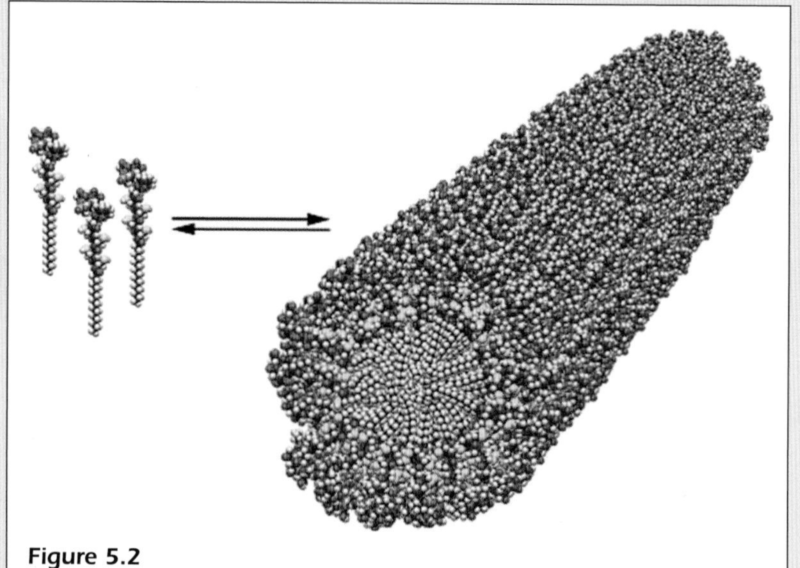

Figure 5.2
Self-assembled molecular template for an artificial bone. The long rod self-assembles from the small molecule components and natural bone tissue forms on the outside edge.
Courtesy of Stupp Group, Northwestern University.

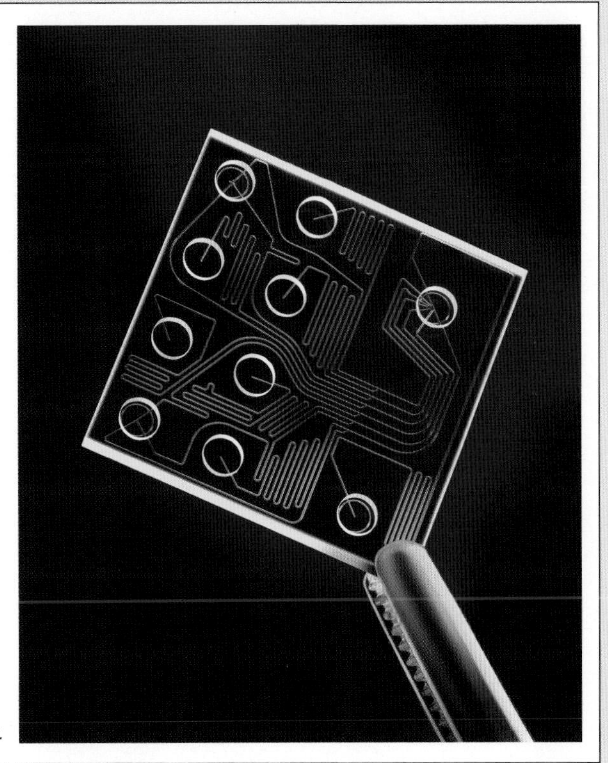

Figure 5.3
A lab-on-a-chip.
Courtesy of Agilent Technologies, Inc.

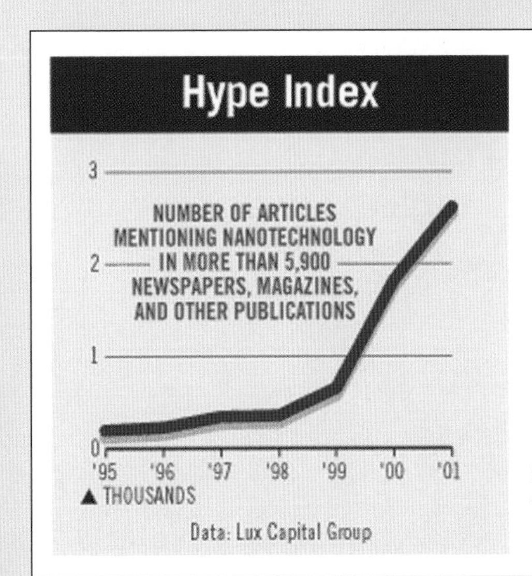

Figure 5.4
The nanotechnology hype index.
Courtesy of Lux Capital.

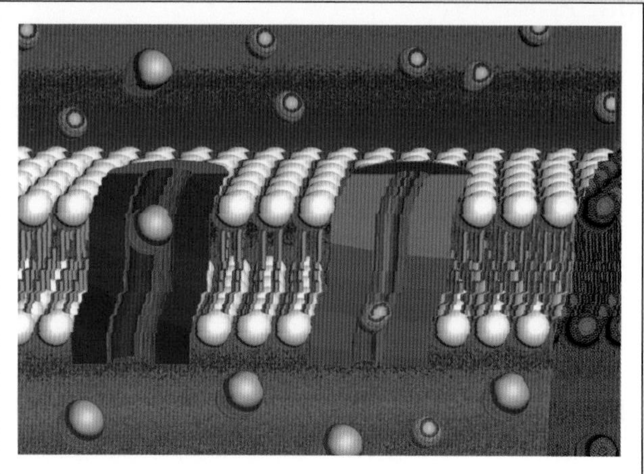

Figure 6.1
A computer-generated model of a portion of cell membrane. The gold balloons are hydrophilic, and the purple thin strands are hydrophobic. The red and blue cylinder structures are channels for moving ions through the membrane.

From General Chemistry, *8/e, by Petrucci/Howard,*

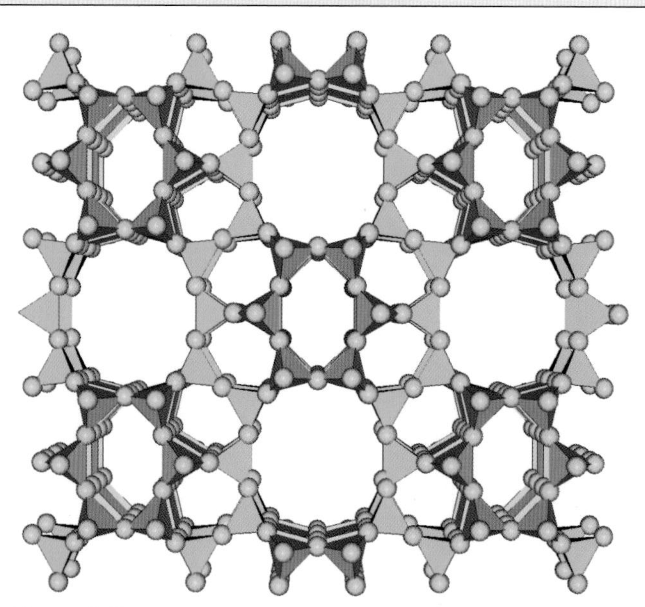

Figure 6.2
A chemical model of a complex zeolite structure. Notice the differently sized holes that represent channels and galleries.

Courtesy of Geoffrey Price, University of Tulsa.

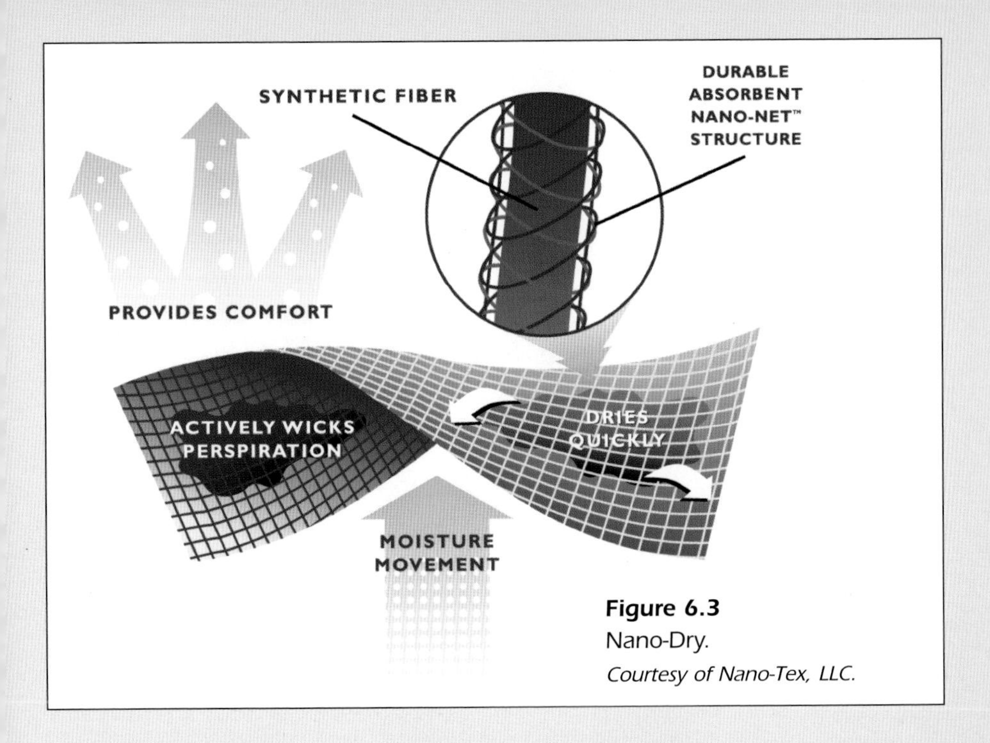

Figure 6.3
Nano-Dry.
Courtesy of Nano-Tex, LLC.

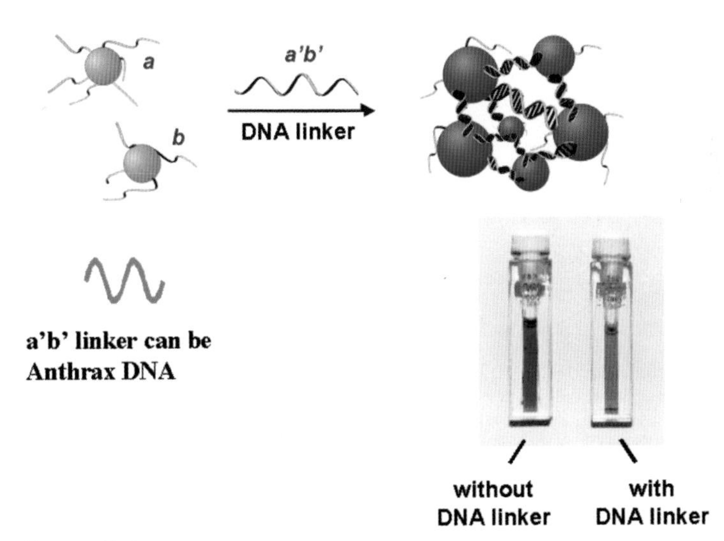

Figure 7.1
The upper schematic shows how the nanodots in a colorimetric sensor are brought together upon binding to the DNA target (in this case anthrax). The clustered dots have a different color than the unclustered ones as is shown in the photograph in the lower right-hand corner.
Courtesy of the Mirkin Group, Northwestern University.

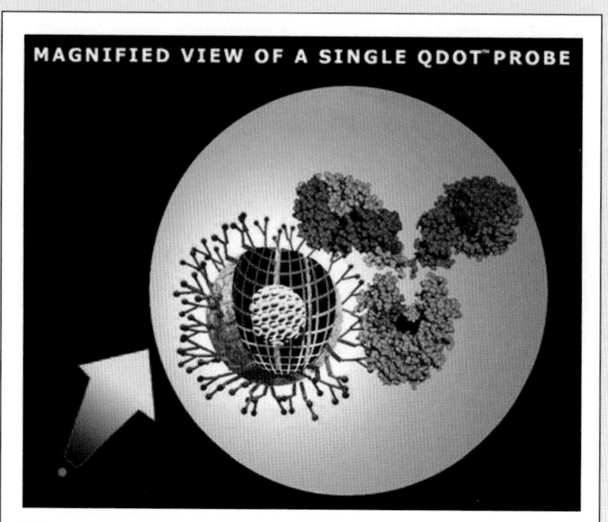

Figure 8.1
Schematic of Qdot probe.
Courtesy of Quantum Dot Corporation.

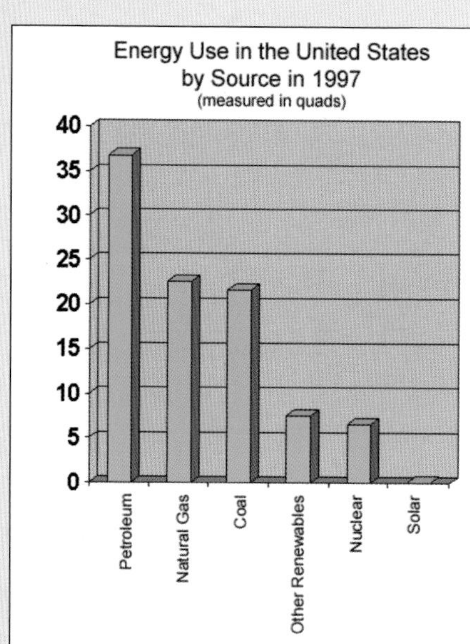

Figure 9.1
Energy use in the United States. If the total amount of solar energy that falls on the United States in one year were added to the graph, the line would be 400 feet long.

the cellular structure. Kinesin is a nanoscale molecular motor that carries molecular cargo through the cell by moving along nanoscale tracks, called microtubules, within the cell. In many respects it is the world's smallest type of train.

Molecular motors are also responsible for signal transduction in the human ear. In this case, the molecular motor is called *prestin* and was originally observed by Peter Dallos and his group at Northwestern.

Molecular motors may provide acceleration and motional energy for artificial nanostructures both within the body and in more complex nanostructure assemblies. Molecular motors are one of the many very complex nanostructures that have evolved through the march of life.

NEURO-ELECTRONIC INTERFACES . . . .

The topics that we have discussed so far are well on their way to actual utilization or have (in some cases) actually been used. In this section we go a bit farther, to discuss one of the visionary aims of nanotechnology. This involves *neuro-electronic interfaces*—the idea of constructing nanodevices that will permit computers to be joined and linked to the nervous system.

The construction of a neuro-electronic interface simply requires the building of a molecular structure that will permit control and detection of nerve impulses by an external computer. The challenge is a combination of computational nanotechnology and bionanotechnology. The nerves in the body convey messages by permitting electrical currents (due to ionic motion) to flow between the brain and the nerve centers throughout the body. The most important ions for these signals are sodium and potassium ions, and they move along sheaths and channels that have evolved specially to permit facile, controllable, rapid ionic motion. This is the mechanism that allows you to feel sensations such as putting your foot in hot water and feeling the heat move from the local nerve through the nervous system into the brain, where they are interpreted and processed. Often this process results in a response being filtered into the muscular system, as is demonstrated when you pull your foot out of the hot water. The aim of neuro-electronic interface technology is to permit the registration, interpretation, and response to these signals to be handled by a computer.

The challenges are substantial: the nanostructures that will provide the interface must be compatible with the body's immune system so that they will be stable within the body for a long time. They must also be able to sense the ionic currents and be able to cause currents to flow backward so that the muscular system can be instructed to perform certain motions. The most obvious such structures will be *molecular conductors*, molecules whose own conduction processes, ionic or electronic, can link with the ionic motion in the nerve fibers.

While active research in this topic is going on in several centers, little practical progress has been reported to date. The wiring up of such a structure is fiendishly difficult because not only must the molecular conductors actually perform the electronic link, but they must also be positioned precisely within the nervous system so that they can monitor and respond to the nervous signals.

The desirability of such structures is enormous. Many diseases involve decay of the nervous system. For example, multiple sclerosis and ALS (Lou Gehrig's disease) both involve nervous system malfunction. Moreover, many traumatic injuries ranging from losing a foot or a finger to nervous system impairment from traffic accidents can result in dysfunctional nervous systems and in conditions as terrible as quadriplegia and paraplegia. If a computer, through a neuro-electronic interface, could control the nervous system, many of these problems could be actively controlled so that the effects of such traumatic injury or disease could be directly overcome by active prosthesis.

PROTEIN ENGINEERING . . . .

Proteins are the molecules that are most obvious in biological systems. Fingernails, hair, skin, blood, muscle, and eyes are all made of protein. Many diseases are protein-based. Some frightening ones, such as mad cow disease and Creuzfeld-Jacob disease, simply arise because a protein is shaped improperly (misfolded). Misfolded proteins also cause several of the genetically inherited diseases, such as Tay-Sachs and sickle cell anemia.

Like DNA and polymers, protein molecules consist of long strings of subunits. In proteins, there are 20 natural subunits that are called amino acids. A given protein consists of a long chain selected from these 20 amino acids, which then assembles itself into a complex folded structure.

Artificial proteins can be made because chemists and biologists have developed methods for stringing together the 20 natural amino acids—and some additional nonnatural amino acids—into the appropriate long strings. *Protein engineering* is the science of making such proteins and then utilizing them both in medicine and in other applications such as synthetic foods.

Protein engineering is one of the nanostructure schemes that underlies an important disruptive technology, sometimes called biotechnology. In its original meaning, biotechnology was the use of synthetic DNA methods to produce particular proteins. Since DNA codes for protein manufacture, biotechnologists hijacked the protein manufacturing mechanisms of simple organisms like *E. coli*, a bacterium found in the gut, but substituted their own artificial DNAs. The proteins could be designed because we know the genetic code that permits a particular DNA to produce a particular protein.

Many proteins have been produced in this way, and some of them have found applications in medicine. A particularly striking case is human growth factor, which is produced by biotechnology and has been used widely in medicine.

Protein engineering is one of the more mature areas of nanobiotechnology because we really know how to make large numbers of proteins. With the mapping of the human genome, the new fields of post-genomic science and proteomics are now being devoted to understanding what proteins do and how their function can be modified or improved by synthetic structures, including entirely artificial proteins.

SHEDDING NEW LIGHT ON CELLS: NANOLUMINESCENT TAGS . . . .

Biologists have any number of reasons to be interested in the movements of particular groups of cells and other structures as they move through the body or even through a sample in a dish. Tracking movement can help them to determine how well drugs are being distributed and how substances are metabolized. But tracking a small group of cells as they move through the body is an essentially impossible task. A needle in a haystack is at least a chunk of dense metal in a mound of light biomass; you can find it by sifting or by using a metal detector. Cells have no significant material differences from one another and are typically of similar size and shape to other cells of their type. Unless you can actually attach something visible and unique to a cell, it will disappear into the body like a grain of sand into a beach.

In the past, scientists got around this problem by dyeing cells. If a sample of cells is green and all the other cells are more or less clear, it's easy to spot the sample. Organic dyes that have been used in the past can be toxic and must still be excited with light of a certain frequency to cause them to fluoresce. More recently, these dyes have been replaced with proteins that naturally fluoresce green or yellow, but these proteins must still be excited by light of the right frequency in order to operate. Different color dyes and proteins absorb different frequencies of light. Consequently, if you have multiple samples that you need to track at the same time, you may need as many light sources as you have samples. This can become quite a problem.

A. Paul Alivisatos, Moungi Bawendi, and their groups addressed this problem with the introduction of what are now often called *luminescent tags*. These tags are quantum dots often attached to proteins to allow them to penetrate cell walls. These quantum dots exhibit the nanoscale property that their color is size-dependent. They can be made out of *bio-inert materials* (materials that do not interact with life processes and are thus nontoxic) and can be made of arbitrary size. This means that if we select sizes where the frequency of light required to make one group of nanodots fluoresce is an even multiple of the frequency required to make another group of tags fluoresce, both can be lit with the same light source. At one stroke, these tags solve two major problems of the old organic dyes: toxicity and the ability to use more than one color of tags at the same time with a single light source.

The science behind luminescent tags is elegantly simple and shows how easily newly discovered properties at the nanoscale can be made

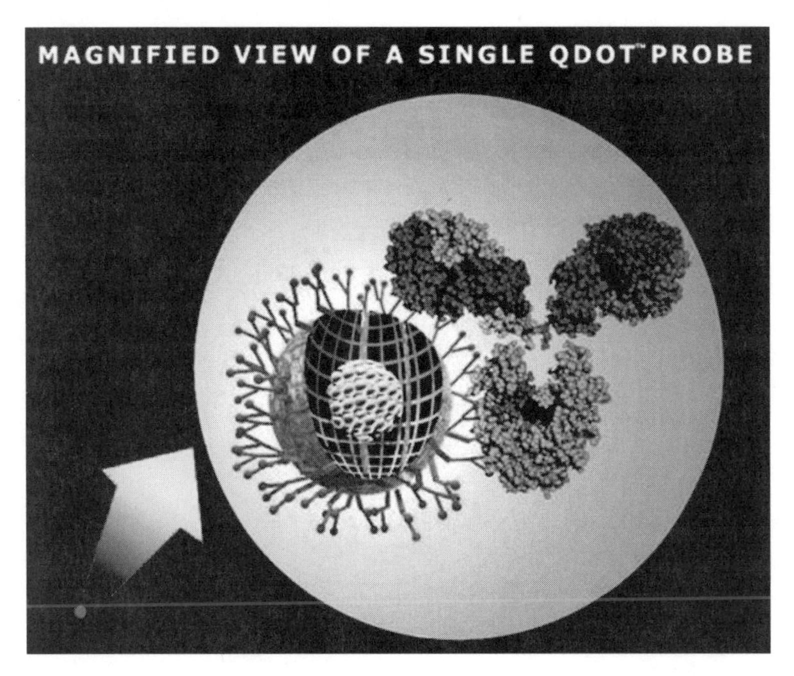

Figure 8.1
Schematic of Qdot probe. *Courtesy of Quantum Dot Corporation.*

practical. Bawendi and Alivisatos have gone on to found Quantum Dot Corporation where this discovery forms the core of their Qdot product. In a Qdot, the quantum dot is surrounded with a shell that protects it from its environment and amplifies its optical properties. The resulting structure can be attached to various different carriers to transport it to whatever it should tag. Figure 8.1 shows how this works.

9 Optics and Electronics

"We have a strategy to reinvent the integrated circuit with molecular rather than semiconductor components."

Stan Williams
Director, Quantum Science Research, HP

In this chapter...

To a great extent the chief driver of the current interest in nanotechnology has been electronics. The electronics industry is facing the likely end of its ability to continue to improve its technology using variations on current chip lithographic techniques, and it is looking for alternatives. Nanotechnology offers several possible solutions to this problem and even allows for the introduction of computers in clothing, wallpaper, and anywhere else, all in communication with the express purpose of making people's lives easier. This application encompasses the notion of *pervasive computing*.

But the implications of nanotechnology for the optics, electronics, and energy businesses also go well beyond better, faster, smaller microchips. Imagine tiny invisible tags that could identify everything from Christmas presents sent through the mail to books in the library to jewelry items, so that their identification and tracking is instantaneous. Imagine efficient solar energy conversion that is really practical and produces renewable energy at less cost than burning fossil fuels. Imagine whole ceilings or walls made of clear or colored cool lights. All these fields are being actively pursued using nanoscience in the general areas of electronic, optical, and magnetic materials. This is perhaps the "highest-tech" area of nanoscience because it includes the interface between nanotechnology and "high technology," or information technology.

Because electronics is in many ways the most obvious application of nanoscience, this chapter is the longest of the four chapters that deal with specific applications. It will be further subdivided into two major sections—the first on light and energy and the second on electronics and magnetics.

LIGHT ENERGY, ITS CAPTURE, AND PHOTOVOLTAICS . . . .

Except for nuclear and geothermal energy, all the energy on earth originally comes from the sun. Oil, coal, natural gas, and other fossil fuels were formed almost entirely by the process of *plant photosynthesis*—plants' method of capturing solar energy. Solar energy also provides for water evaporation, and therefore indirectly for rain and hydropower. It also continues to warm the world, powering the winds that give us wind power and the production of the biomass that provides food for all creatures. Thus it is clear that light is the major source of energy as well as a means of communication, data storage, and information display. Nanostructures are important for all of these applications.

Natural photosynthesis occurs in many different organisms. The most obvious and significant are plants, which use solar energy both for the synthesis of the starchy material that they are made of and for production of oxygen. Photosynthesis therefore not only gives us our energy but also the very air that we breathe.

The natural photosynthetic apparatus is an enormously complex, beautifully designed set of intertwining nanostructures. Actually, the photosynthetic centers in bacteria are more completely understood than those in green plants. In these bacterial structures, several contributing nanocomponents are held together with membranes in the usual way that cells are constructed. Photosynthesis works by using the energy from sunlight to separate positive and negative charges, which are then transduced into proton charge gradients. The recombination of these charges produces energy.

Three major nanostructures are involved in this process: the antenna, the reaction center, and the membrane charge management structure. The antenna consists of a large number of molecular light-absorbing centers, each of which absorbs energy from the sunlight and moves that energy into a collecting station subcomponent called a *light-harvesting complex*. This consists of rings of individual large molecules that exchange the energy among themselves until the energy is ready to be transferred into the next nanostructure down the line, the reaction center. The reaction center is where the energy is used to separate an electron from the opposite charge (called a hole). The electron is pushed quite a long way from the hole (they are separated by more than 2 nanometers) and is captured by another nanostructure consisting of an iron atom and a few organic molecules called *quinones*. The final step involves utilizing this transferred charge in an electronic system to generate a gradient of hydrogen ions across the membrane. Very complex membrane-bound structures then permit charge recombination, eventually resulting in the formation of ATP, the molecule that is fundamental to storing and carrying energy throughout biology.

This set of elegant and beautifully constructed complex nanostructures effectively captures energy by performing three functions. First, the sunlight is captured; second, the sunlight is used to separate positive and negative charges; and third, the positive and negative charges are recombined in such a way that the Coulombic energy of combination is filtered off and used to create a useful energy source.

Constructing synthetic devices to capture sunlight to produce energy has been an enduring scientific dream. Schemes from boiling water to melting snow to warming stones have been employed in the past. The most attractive ideas, however, involve constructing devices to transform the energy from the sun directly into electricity or into a form of stored chemical power, such as the hydrogen molecules that can be obtained by splitting water using solar energy or even like ATP. All these chemical storage schemes can be efficient for energy generation, but they do require an additional conversion step to create useful electrical power (as with fossil fuels, which are a form of chemically stored solar energy). Consequently, most of the interest has been in the area of *photovoltaics*, devices that directly convert sunlight into electrical energy.

Photovoltaic solar energy conversion research and development has focused largely on utilization of semiconductors, particularly silicon. Silicon photovoltaic cells are found in residential and industrial construction, in toys, in remote sites that require contained power generation, and (perhaps most visibly) in hand-held calculators that use light energy. These structures mimic natural photosynthesis, but only part of it. There is no antenna structure. Rather, sunlight focuses on a semiconductor (normally silicon single crystals or polycrystalline silicon, although other semiconductors are also used). The semiconductor absorbs the energy, and the excited states of the semiconductor permit the electron and the hole to be separated to opposite sides of the power cell. There the electrons and holes are drawn to so-called current collectors, and the difference in their energies is the result of the energy absorption from the light source. The electrons and holes are then allowed to recombine by passing the electrons along a wire and generating an electrical current. That electrical current can be used to power a home or a calculator, or it can be sold to the power grid.

Many economic as well as scientific considerations determine the practicality of solar photoconversion. Some factors include the amount of sunlight falling in a particular area, the aesthetics of solar panel construction, and the prevailing cost of electricity and its availability. More scientific issues include the efficiency of the solar cell, defined in terms of the amount of electrical energy that is in fact derived from a given amount of initial light energy from the light source (energy out over energy in). Other factors include the manufacturing cost of the device, eventual disposal cost, and any maintenance or toxicity issues. The driving concern so far in advanced countries like the United States is the initial cost of such solar cells; therefore, extensive research is being conducted to reduce this cost. Until mass production is possible, however, the price barrier will be hard to break, and this is likely to happen only if solar power becomes a priority of public policy and there are preorders to fill factories. Still, nanotechnology breakthroughs will make a tremendous difference.

Natural photosynthesis, of course, is not done with semiconductor crystals but with molecules. Therefore, one major nanoscience approach to the problem of improving solar conversion practicality is what is sometimes called *artificial photosynthesis*, using nanostructures based on molecules to capture light and separate positive and negative charges. The molecular nanostructures used for this purpose come in several different flavors. In the simplest situation, the molecule has only one working part that actually absorbs energy from the light. The molecular excited state then falls apart, with the electron directed toward one electrode and the hole moving toward the other. Such structures are relatively simple, but often the efficiency is not very good, both because absorption intensity is limited and because the charge separation into the electrodes can be highly inefficient.

More complex structures involve what are called dyads, triads, or even pentads. This is a Greek way of saying that the molecular nanostructure has several subunits, one of which captures the sunlight by absorption, while others are instrumental in facilitating an efficient separation of electron and hole.

Molecular nanostructures have some advantages compared to semiconductor-based structures, including lower cost, lighter weight, and fewer issues with toxicity and environmental aspects. So far, however, their relative efficiencies are much lower. This is true because of both the inefficient capture of sunlight and the losses of energy and efficiency in the conversion from the excited states to the separated electron and hole that eventually combine to give current. Still, artificial photosynthesis remains a promising and active direction of research, as are semiconductor solar cells and hybrids like the Graetzel cells mentioned in Chapter 5. In that hybrid approach to nanoscale photovoltaic material, an organic molecule (the dye that originally absorbs light) is combined with a nanostructured electrode made of titanium dioxide (a semiconductor), and the result is efficient charge separation. Just such hybrid structures between soft (molecular) and hard (semiconductor) nanostructures may permit substantial advantages in terms of stability, efficiency, and cost.

Solar power will remain a key focus of nanotechnology because its promise of clean, renewable energy is attractive for economic, political, ecological, and societal reasons. This can be quantified to some extent. The total energy contained in the sunlight that shines on the 48 contiguous states of the United States in one year is around 45,000 quads, and the total amount of energy of all types that the United States uses in one year was approximately 94 quads in 1997, or 0.2 percent of the solar potential. A quad is a unit of energy that is almost unimaginably large—it is roughly a million billion BTUs. Figure 9.1 shows other energy numbers in comparison with this number. Note that the United States is a major producer and consumer of energy, and that almost all energy worldwide is still obtained from fossil fuels. Renewable energy sources and nuclear power still make only a minor contribution though they could easily power everything. This is a major challenge for nanotechnology.

LIGHT PRODUCTION

In photovoltaics, the aim is to use sunlight to produce chemical or electrical energy. In light emission processes, exactly the opposite happens: either chemical energy or electrical energy is used to produce light. Nature shows many examples of light production such as luminescent bacteria and organisms like fireflies that utilize particular molecular structures to produce light.

Of course electric light and fluorescent lights are common, and have been for nearly a century. In fluorescent lights, molecules are

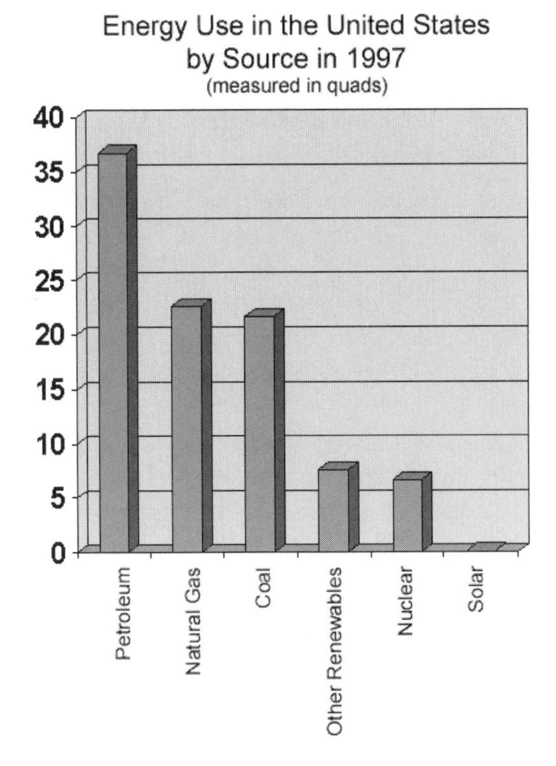

Figure 9.1
Energy use in the United States. If the total amount of solar energy that falls on the United States in one year were added to the graph, the line would be 400 feet long.

excited by collisions with electrons that pass through the fluorescent tube; these excited molecules then luminesce in exactly the same way that fireflies do. In incandescent lights, a wire is heated to a very high temperature, where it emits both light and heat. (The heat is wasted energy and is certainly not desired, especially by people who brush past a bare bulb.)

Nanoscience has entered the field of light emission largely in connection with so-called light emitting diodes. A light emitting diode is exactly the opposite of a photovoltaic. In LEDs, the oppositely charged electrical carriers, the electron and the hole, recombine to form an excited state. That neutral excited state then loses its energy by light emission. This last process is very much like the fluorescent tubes, but the way in which the molecules are excited is quite different. In fluorescent lights, the molecules are excited by collision with rapidly moving particles. In LEDs, they are excited by recombination of electron and hole charge that provides enough energy to excite the molecule or semiconductor locally—the excited structure then emits visible or infrared light (and a lot less waste heat).

Most LEDs are based on semiconductors, just like most photovoltaics. The relative advantages of semiconductor and molecular structures in light emission are precisely the same as those in photovoltaics; however, the efficiency of such molecular devices is still substantially lower than that of the semiconductors. The advantages of molecular structures, including environmental safety and cost, put both approaches (and perhaps hybrids) under active pursuit. As noted in Chapter 5, Ching Tang and his colleagues at Kodak first introduced molecular LEDs. Reducing them to nanoscale size is the work of Tobin Marks, S. T. Ho, and their groups at Northwestern.

Utilizing luminescence is also of substantial interest in a number of applications. For example, luminescent bar code structures are available. These consist of multiple sites along an extended nanostructure that luminesce selectively when responding to a particular signal. Such bar codes can be used both as sensors and as tags.

Recent work from Charles Lieber's group at Harvard has demonstrated that crossed wires, made of semiconductors with nanoscale dimension, can act as light-emitting structures. These crossed-wire emitters are probably the smallest current light sources; they are intense and their colors can be chosen. Along with the rest of the LEDs, these nano sources are very promising for applications ranging from full room lighting to extremely high-resolution displays. Flat screen, high-resolution, very bright, and very efficient LED structures are beginning to appear in specialty niches such as military vision systems, cellular phone displays, and even dashboard controls on automobiles. Light emission, like photovoltaics, is one of the most immediately attractive applications of nanostructures.

LIGHT TRANSMISSION

Communications are crucial for modern society. Thirty years ago there were no cell phones, no Internet, no World Wide Web, no discount long-distance carriers (and many fewer telemarketers). Vastly improved communications networks resulted in making the world seem smaller and more intimate, and (in an optimistic view) closer and less warlike.

Almost all modern communications involve the transfer of messages using different parts of the electromagnetic spectrum. There are long wavelength radio stations, shorter wavelength microwave towers, and very short wavelength optical communications. Optical fibers have been one of the real technological successes of the past two decades. Fiber optics permit high-speed, efficient, high-density, high-reliability passage of enormous densities of signal. A single-fiber network can carry tens of thousands of data streams and voice conversations at once. The capacity of these fibers has not reached any intrinsic limit, but it is checked by the capacity of the electronics at either end.

Nanoscience and nanotechnology have been active in producing effective fiber-optic structures. Fiber-optic cable is actually a fairly complex structure because the inner shell material should differ from the cladding so that the signal can proceed without too much scattering or destruction over long lengths (now as long as thousands of miles without requiring external amplifiers or repeaters). The inner fiber itself should be as free from scattering loss—inhomogeneities, cracks, impurities—as possible and may even be doped with nanostructures to recover diminishing signals; the signals may be polarized with nanostructures for better multiplexing. Because light can be scattered by relatively small particles, synthetic methodologies for reducing impurities in fiber optics are a major industrial ambition. Nanotechnology is addressing all these problems.

LIGHT CONTROL AND MANIPULATION . . . .

In current technologies that pass information along optical fibers, it is necessary to transform the information from optical into electronic signals for actual manipulation of the signal, including routing, switching, and processing. If optical components could be constructed to reliably switch, amplify, route, and modulate optical signals, then it would be possible to construct an all-optical communications system. Moreover, if two light signals could be controllably combined to give a third one, we can envision a full computational scheme based only on light. In such a scheme, the optical computer would store its information on an optical disk or in a holographic or optical crystal, and the manipulation of data, actual computations, would be carried out using light rather than electrons.

The obstacle here is the construction of small (ideally nanoscale) optical devices that in fact allow manipulation of light signals. Materials should have particular properties to allow optical switching—the control of light by light—to exist. These materials were predicted and demonstrated as early as the 1950s. The current difficulty with such materials, referred to as nonlinear optical materials or NLO structures, is that their efficiencies are very low and their characteristic sizes are still far too large. As we will see in our discussion of electronics later in this chapter, Moore's law is driving electronic computational components toward the nanoscale very rapidly. Most optical switches are still big enough to be seen by moonlight with the naked eye.

As we have seen so often, the appropriate materials can be based on semiconductors or on molecular entities. The best NLO performance currently is found in crystalline materials, particularly those based on lithium niobate. These are single-crystal devices, and as such are fragile and have particular assembly challenges. Nevertheless, modulators based on lithium niobate are incorporated into several current optical systems, and miniaturization of such structures is proceeding rapidly.

For the usual reasons of assembly, cost, size, and safety, molecular-based NLO structures are of substantial interest, but so far this remains another area in which molecular species do not yet match the performance of semiconductors. So-called third-order nonlinear optical materials are necessary for optical computing and optical data manipulation to ever become realistic. In this area, neither semiconductor nor molecular technology has yet advanced to the point where devices are practical. This lack of efficiency arises from the nature of NLO response, which depends on the intensity of the light of two, three, or even four incident beams. Because the interaction of matter and light is inherently relatively weak, response to several beams at the same time is an event of low probability; therefore, the

NLO response of materials is generally weak and inefficient. Schemes to improve this are abundant throughout the scientific community, and this is once again an area of great challenge and potential applicability.

Optical transport and switching are subject to quite different constraints than electronic processes because the photons that carry light do not have a charge. Because electrons do have charges, they can interact with charged ions as they move through wires. These interactions are relatively weak, and the conduction in perfect metals can be very efficient. When the metals are less pure or have structural defects, the electrons move less efficiently and scatter, depositing energy in the form of heat into the material. This resistive heating is often useful—it is how toasters operate. In electronic circuits, however, the resistive heating causes dissipation of energy, wasted power, and sometimes catastrophic failure. Anyone who has worked with a laptop on bare knees recognizes the hot problem of resistive power loss. In addition, high-frequency electronic devices such as computer networks and microprocessors face issues of mutual and self-inductance: circuits can act like antennae and signals can jump from wire to wire when frequencies are high and wire spacing is low. Making wires coaxial or twisting pairs reduces this problem in computer network technologies such as Ethernet, but there is no technology to do this on an integrated circuit. As chip clock speed and density increase, these issues of inductance and resistive heating are becoming major roadblocks. Optical computers and optical devices, because they do not move electronic charge, are nearly immune to both problems. This is one of the signal advantages of optical circuitry.

ELECTRONICS . . . .

Electronics is currently the workhorse technology for computing and communications as well as a major component of consumer goods. Though few of us are old enough to remember crystal radios, some can remember vacuum tube electronics. When we first talked about the advent of the transistor and then of integrated circuit and silicon chip manufacture, we asserted that these were powerful economic, societal, and technological forces and that their development led to the growth of high-tech applications, which has dominated the industrial and commercial progress in the developed world in the last third of the 20th century. The ongoing development of electronics continues to provide a huge bounty both financially and in terms of quality of life. We also said that we are unlikely to continue at anything like our current rate of progress in electronics development unless there is a major technological revolution in the way electronics work and are made. That revolution may be made possible by nanotechnology.

To maintain focus, we will limit our discussion of electronics largely to chip-based structures since the current technologies surrounding them are those most rapidly reaching their limits. We will also say something about memory and interconnect structures that are needed for efficient computation. In making this choice, we intentionally will ignore the tremendous applicability of nanoscale electronics in many communications and consumer marketplaces, as well as in other areas ranging from radar to radios to routers. In keeping with the theme of this entire book, we will focus on some of the most challenging and promising areas. There, as we've already stressed, nanoscience and nanotechnology will pervade all aspects of our lives for the next several decades.

At first sight, the Moore's law graph in Figure 2.4 seems absolutely smooth, suggesting that advances in electronics development have been continuous. Actually, they have been episodic. A large number of improvements have been developed by creative engineers, allowing chip-based technology to become cheaper, denser, and more efficient. Still, several fundamental issues suggest that there is a brick wall ahead that will block this ongoing improvement. This wall will arise from some fundamental physical limits due to the nature of electrical conduction and the requirement that a transistor must be able to be turned on or off by voltage across its gate. When the transistor gets too small, quantum mechanical leaking of the electron through the transistor will mean that it is no longer clear whether the transistor is supposed to be on or off. This will call for entirely new logic approaches or, perhaps, for different nanoscale structures.

Advanced technology within the semiconductor industry is already at the nanoscale. A 130-nanometer technology is becoming standard in current chips, and advanced prototype models that reduce the size substantially further are available. As we have already indicated, manufacturing such nanosize objects using the top-down lithographic techniques that all semiconductor fabrication facilities now utilize is driving up costs at an exponential rate. Given the increased cost and the remarkable tolerances necessary to continue Moore's law, other schemes for electronics become attractive.

CARBON NANOTUBES

The carbon nanotube structure has already made its appearance several times because it represents an entirely new form of matter. Single-walled nanotubes can be either semiconductors or metallic. Nanotubes are also very stiff and very stable and can be built with their length exceeding thickness thousands of times.

Nanotubes can, however, exhibit even more interesting behavior. Scientists such as Cees Dekker in Delft, Paul McEuen at Cornell, Phaedon Avouris at IBM, and Charles Lieber at Harvard have demonstrated that single nanotubes can actually act as transistors. Pairs of nanotubes, or crossed nanotubes, have been shown to work as logic structures. These experiments constitute a proof of principle that nanotube logic, at an unprecedentedly small scale, can actually provide a modality for computation.

The fundamental science of nanotubes is very exciting, and many of the major academic efforts in nanotechnology center on carbon nanotubes, with important efforts at places like Rice, Harvard, Cornell, Northwestern, Tsukuba, Delft, Tokyo, Stanford, Georgia Tech, Illinois, North Carolina State, and Cal Tech. One of the major challenges of nanotubes has been physical assembly. Because nanotubes tend to stick to one another and do not exhibit the molecular recognition properties more generally associated with organic molecules, manipulating nanotubes using bottom-up techniques unaided by molecular recognition remains a major challenge. Hybrid structures between nanotubes with their attractive physical and electrical properties and soft molecules with their assembly and recognition properties constitute an attractive avenue toward the construction of electronic devices based on nanotube function.

SOFT MOLECULE ELECTRONICS . . . .

Using more traditional organic and organo-metallic molecules as electronic components offers some aspects that are more attractive than using nanotubes, including both relative ease of assembly (and potential for self-assembly) and some of the control and recognition (including biorecognition) features that molecules permit. While most organic molecules are soft insulators—think of wax and polystyrene, tar and fingernails—under particular conditions these molecules can conduct current. Indeed, current transport in molecules can be controlled either by chemistry or by electromagnetic fields.

The advent of scanning tunneling microscopes described in Chapter 4 has led to burgeoning interest and activity in the field of molecular electronics. Within the past two years, scientists have demonstrated that single molecules can switch like transistors, that molecules exhibit nondissipative passage of electrical current (effectively superconduction, but via a different mechanism—this is something not previously demonstrated in traditional integrated circuits of any size), that molecular structures can be true superconductors, and that molecules can be used as active switches in electronic circuits. This remarkable set of discoveries has placed new emphasis on the possibility of using molecules as components in electronic devices. These possible applications range from molecular interconnects or wires through molecular switches to molecular insulators, molecular assemblers, and molecular memories. This would lead to the highest-density computers possible using current computer architectures. Efficient assembly of these devices is now perhaps the greatest challenge of molecular electronics.

MEMORIES . . . .

In constructing a computer, or any other electronic device, it is important to store information on a temporary or a long-term basis. Information storage is done in memory devices, and different sorts of memories have been used since the introduction of the early magnetic core memories. Indeed, the increase in the capability of memory—

that is the amount of information that can be stored in a given volume of space—has improved even faster than the Moore's law curve for transistor density.

The current state of the art is represented by hard-disk memories. These are based on magnetics: information is stored as magnetic polarization on a disk and is read or written by a special head as the disk spins. The phenomenon involved here is called *giant magnetoresistance,* which refers to the effect of the magnetic fields on electrical resistance. Depending on the magnetic polarization (whether the bit of information is a one or a zero), the currents registered in the read head will differ. The giant magnetoresistance phenomenon has made the disk business an enormous one—now roughly $40 billion per year worldwide.

Using nanostructures, it is possible to reduce the size of memory bits substantially more and thereby to increase the density of magnetic memory, increase its efficiency, and lower its cost. Chris Murray's work at IBM (see Chapter 5) is typical of the best research being done in this area. In Murray's work, the bits are stored as magnetic nanodots; these dots can be made very precisely, and their size can be reduced until it reaches something called the super paramagnetic limit. For smaller dot sizes, the magnetic storage is not stable, and it can be interfered with by thermal energies. Therefore, the super paramagnetic limit represents the smallest possible magnetic memory structure.

The nanolithographic methods discussed extensively in Chapters 4 and 5 are already being used to prepare some strikingly powerful memories. For example, using a soft lithography method such as dip-pen nanolithography, nanocontact printing, or controlled feedback lithography, it is possible to reduce individual features down to the size of a few nanometers. If such dots each contained one bit of information (either a zero or a one), and if they were spaced by distances equal to ten times their own sizes, then on a piece of paper as large as the one you are reading, you could easily store 100,000 sets of the *Encyclopedia Britannica*.

Nanotechnology and nanoscience offer different memory possibilities. For example, photorefractive materials were also discussed in Chapter 5. Such materials represent only one kind of optical memory. CDs and DVDs that are used to record music and movies also use a type of optical technology, with the reading done by lasers.

Current magnetic memories and optical memories are largely two-dimensional; they are based on a flat surface. Memories such as holographic memories and photorefractive memories are based on the interaction of light with matter. In such memories, information is stored by changing molecular states with high-intensity laser fields. Lasers are used to write information into the memory, information that can be changed by further high-intensity laser radiation, or can be read by low-intensity light. One of the striking advantages of such nanoscale optical structures is that they can exist in three dimensions because not only the surface but also the bulk of the material is being read. This could lead to far higher efficiencies and storage capabilities for optical memories.

We can see then that in memories, just as in logic and computing devices, there are advantages to using optics as opposed to electronics. Once again, the problem so far is that these very sophisticated optical techniques require both materials properties and instrumental capabilities that are not assured. Development of optical materials to use as memories will be easier than optical materials to use as switching devices because both the fabrication and the reading depend on linear processes, processes that are proportional to the intensity of one light beam. Optical memories that are already used will be developed at the nanoscale before any truly nanoscale optical logic structures, but both nanoscale devices will provide exponentially greater capacities if they operate in three dimensions.

Other methods for very high-density memories have been proposed in the general area of molecular electronics. DNA memories were discussed in the sidebar on DNA computing: nature uses DNA to store its genetic information, and computational information can also be stored in DNA structures. There are several problems with this, including speed, the method of read-out, and the fidelity of the DNA structure, which cannot be prepared using standard lithographic techniques in any obvious way. Nevertheless, the promise of inexpensive, high-density, reliable DNA-based memory and the prospect of massively parallel DNA computing will make this field interesting.

In connection with both ordinary memory and quantum computing, molecular conductors are of substantial interest. In particular, the so-called magnetic "spin valve" structure can be used to integrate electronic motion in molecular structures with memory. In these structures, the ability of an electronic current to pass through a molecule depends on the spin of the electrons (remember spin from the sidebar on quantum computing?). By changing the local molecular structure, it is possible to permit charge of one spin to pass through, while charge with the opposite spin is blocked. Both the Cornell group (Hector Abruña, Paul McEuen, and Dan Ralph) and Hongkun Park's group at Harvard have observed such spin-dependent transport. Ron Naaman's group at the Weizmann Institute in Israel has shown the capability of individual molecules to perform many tasks, including differential transmission capabilities for oppositely polarized electrons. These developments could lead to memories where a bit is stored on a single molecule.

GATES AND SWITCHES . . . .

Standard computer and microelectronic design is based on the use of the *field effect transistor*, which is a very simple switch that can be turned from on to off by applying voltage to its control (called a gate) electrode. By combining and cascading these transistor on-off switches, it is possible to create higher logic structures called *logic gates* (not related to the gate electrode, which is the control interface for a transistor, or to Bill Gates, who occasionally defies logic. Logic gates are also known as Boolean logic gates, after George Boole, a 19th-Century English mathematician). Logic gates can perform a variety of logical functions on input signals including AND, OR, and NOT. Combining logic gates with memory devices allows us to build processors and all the other core parts of a modern computer system. For this reason, transistors remain the heart of digital computing. We have already indicated in our discussion of nanotubes that nanoscale structures can act as transistors, and we have remarked on the importance of work demonstrating that individual molecules can act as field effect transistors, with sizes more than 100 times smaller than those afforded by currently available standard lithographic work in silicon.

A transistor is naturally in one state, on or off (which one depends on the type of transistor). Pushing it to the other state requires the application of voltage, which must be maintained as long as the other state is required. As soon as the voltage is removed, the transistor drops back to its natural state. This is similar to a coffee grinder where you need to hold down the button for as long as you want the coffee to grind. Other components, simply called switches, are stable in either state. They will stay on until flipped off or off until flipped back on. You need to apply voltage only when you want them to flip between states. This property is called being *bistable*. Molecular structures have been used to provide switching capabilities, this time based on the concept of their dynamical states, where molecules with the same bonding capability can nevertheless have two different physical structures, sort of like the open and closed forms of an umbrella. In particular, a collaboration between Jim Heath's and Fraser Stoddart's groups at U.C.L.A. and Stan Williams's group at Hewlett Packard has demonstrated that molecules called rotaxanes can be used to switch the current passing through a molecular wire from a high current state to a low current state. This switching is the basis for an entire integrated architecture—that is, the switches are placed in an array of logic gates that can actually perform computational tasks.

ARCHITECTURES

If computers could only answer very limited sets of questions such as comparing two pieces of information (as in a logic gate), cracking a code (as in quantum computing), searching for data (as in DNA computing), or even solving complex mathematical problems (as in swarm computing), they would not have changed modern life. Computers are general-purpose; they can turn on a coffee pot in the morning, calculate income tax (inasmuch as this is possible), guide a jet airplane through a thunderstorm, and, through CAD tools, allow an engineer to design yet more advanced systems. The overall structure of a computer's design—its architecture—is what suits it to a particular application or allows it the versatility to be used for many applications.

So DNA, quantum, and swarm computing are very powerful, but they have limited applicability. Gates alone are too simple to perform useful work no matter how fast, small, efficient, and sophisticated they are. But the approaches that we've discussed in this chapter—all-optical computing, molecular electronics, and nanotube-based circuits—could operate using the same basic designs and architectures as current computers and would be general-purpose, allowing them to be simple replacements for current systems.

One of the great advances that has permitted Moore's law to continue is our increased capability, using silicon lithography, to align and connect a large number of logic gates, producing arbitrarily complex architectures for general digital computing capability. But the limitations inherent in top-down lithography place a limit on the kinds of architectures that can be developed using current CMOS methods. The accuracy that a pattern can reproduce depends on the wavelength of the light used in that lithographic pattern; even light far into the ultraviolet has wavelengths on the order of 10 nanometers. X-ray lithography could be done with smaller wavelengths, but x-rays have high energies (energy is proportional to frequency and inversely proportional to wavelength) so that they can damage materials as well as create them. Also, such lithography is generally limited to a single plane or a series of planes.

Some of the soft lithographic methods mentioned in Chapters 4 and 5 can be used, in a bottom-up scheme, to produce complex arrays of small features at relatively low prices compared to CMOS prices. The capabilities of such lithographies to make extended structures are under current development in many places. The use of soft lithographies and molecular-based computational structures is a major intellectual and engineering challenge, but the inherent elegance and power of bottom-up design, assembling things atom by atom and molecule by molecule just as nature does, is one of the most exciting aspects of nanoscience or of any science today.

10 Nanobusiness

In this chapter...

BOOM, BUST, AND NANOTECHNOLOGY: THE NEXT INDUSTRIAL REVOLUTION? . . . .

After seeing the potential of nanotechnology, it's easy to get euphoric. Everyone is looking for "the next big thing" to bring back economic good times, and nano is becoming the betting person's favorite. Although it is not yet clear which of the many technologies we've discussed will actually turn to gold (of any color), it is certain that many of them will have a substantial impact. The National Science Foundation estimates that the industry could grow from essentially nothing to $1 trillion worldwide in just fifteen years, a dizzying level of growth. In the end, nano would be a bigger economic force than software, cosmetics, drugs, or automobiles are in the United States today—almost bigger than all of them combined.

This is a slightly deceptive way to look at the sector. It is based upon growth projections made with little industry data, and it is likely that a large percentage of the economic growth emerging from nanotechnology will remain within existing market sectors such as industrial chemicals and energy, as established companies go nano. Since inventions like high-bouncing tennis balls and stain-proof pants that employ nanotechnology may not be marketed as nano, it may be difficult for analysts to find the nano in the stew of industry, but it will be there, adding its spice to the economic whole.

Not all the potential of nanotechnology lies in established companies and established sectors. To capitalize on the potential of the new market, the temptation to form start-up companies will be strong. Investors and entrepreneurs can make (or lose) much more money growing enterprises from scratch than they can by betting on the incremental growth of larger companies. In most cases, emerging businesses represent the only route to three- or four-figure returns on investment so even after the recent downturn, start-ups and spin-offs are seen as the way to win big. But after the boom and bust of the high-tech market from 1998 to 2001, we've seen that this is a risky path, and it is important to think about how the industry will evolve and do some planning now, while nano remains nascent.

Few nanotechnology products have yet appeared on store shelves. For the most part, companies currently pursuing nanotechnology initiatives are involved in pure research and development or in making raw materials for other people's nanotechnology research or industrial production—things like zeolites, nanowires, and custom genetic material. Private investment in nanotechnology is still a trickle, though it is rapidly showing the signs of becoming a flood. The industry is still very much in the stage of formation with a golden future if it is managed properly and the lessons of the tech boom are learned. But if euphoria and hype again exceed proper scrutiny, diligence, and business principles, there is even more scope for investors, entrepreneurs, companies, and employees in the sector to be fooled and hurt. If this happens, real, sustainable growth and innovation will slow, and, as people become desperate, there is even greater risk of unethical applications and distribution than there is with information technology.

We have not had much time to learn from our mistakes, but some are already clear. Nanotechnology does have some features that may allow it to avoid many of these problems, but it also has many new challenges of its own. In this chapter, we'll look at the current state of nanotechnology business as it enters the starting gate and sweep the track ahead for some of the potential pitfalls. We'll also look at some of the unique potential of the nanotechnology business and see why its emergence doesn't need to follow the rocky road of the dot com era.

NANOBUSINESS TODAY

The landscape of nanotechnology is already being split among three kinds of entities. First are the open research labs including universities, national laboratories, and programs within government agencies such as National Institutes of Standards and Technology (NIST) and the National Institutes of Health (NIH). Second are large corporations with research, development, manufacturing, marketing, and distribution capabilities such as Merck, IBM, Dow, Kraft, 3M, and Agilent (formerly part of Hewlett Packard), and third are the start-ups and spin-offs formed by professors, researchers, graduating students, and others who have seen an idea in a laboratory and want to commercialize it.

Nanoscience is a moving target; there are many directions in which research can go. Commercial need for specific products will certainly nudge research to move in certain directions, but there is still so much to be understood about the fundamental behavior of nanostructures that whole new areas of interest could emerge. A large amount of money is needed for basic groundwork that may or may not result in commercial products. Universities and government institutions exist to do this kind of pure scientific research. It is important that governments and corporate donors around the world actively promote this and that institutions create cross-disciplinary centers for nanotechnology. There are already several key centers of this type in the United States. They include Northwestern, Harvard, MIT, Rice, Illinois, Purdue, Cornell, UCLA, Texas, and Berkeley. Unlike silicon technology, nanotechnology isn't rooted at the coasts.

Many universities are equipped for nanoscience research; however, even for them, it will not be business as usual. Not only will they require specialized centers for nanoscience, but they may also face the "brain drain" that characterized the dot com bubble at places like MIT and Stanford. Top researchers will find themselves with discoveries of great commercial value and will be put in a position of having to choose whether to launch enterprises, license their discoveries, put them in the public domain, or put them in the trust of their institutions. These researchers and their students are already being seen as the technology transfer agents of nanotechnology, and they will be given huge compensation offers if they move to the private sector. This might make it very hard for universities to retain them. Unlike information technology, where almost anyone can train to be a technician and get to the point of being productive in a matter of months, nanoscience research will, in most cases, require in-depth scientific knowledge and a PhD-level background. In the short term, there is likely to be a shortage of talent, especially if nanotechnology start-ups become more common.

Big corporations will have some advantages in the world of nanotechnology. Many of them can already fit nanotechnology into existing product lines, and they will prevail where manufacturing and distribution can be expensive and there are massive economies of scale. Some obvious winners will be companies in the pharmaceuticals and microchip industries. Pharmaceutical companies have the clout to get drugs through the Food and Drug Administration (FDA) approval and the credibility, channels, money, and legal wherewithal to get them to market. Microchip makers have billion-dollar fabrication facilities, things beyond the budget of even the best-funded start-ups.

Large corporations also face the same challenges that they did during the technology boom. They have slow and cumbersome processes for approval, adoption, and commercialization of new products. A new idea must be sold to executives within the company before it can be sold to customers outside the company. Overburdened sales and marketing departments may not be able to adapt quickly to large changes in company product lines, and new products can get lost in favor of less up-to-date cash cows. Also, most large corporations are public and, after the unforeseen collapse of giants like Global Crossing and Enron, are coming under increased scrutiny especially when it comes to gambling on new technologies. Smaller, private companies are typically more closely held and less risk averse. To some extent, the big, global corporations may have to rely on these smaller concerns to test the waters and then license successful technologies or buy the enterprises that spawned them. But let the buyer beware! Over the last few years Lucent Technologies and Cisco were not alone in getting themselves into trouble making unwise (and overpriced) purchases of companies that had ephemeral claims for their products that did not really fit into any core product lines.

Start-up companies will fill the remaining part of the nanotechnology landscape. At this time, the very words "start-up company" are sometimes enough to get you quietly blackballed from polite conversation or sent out to play with the other children, but this is unfair. The concept of start-ups remains sound, despite the abuse it has received in the last few years. With the benefit of 20/20 hindsight, many of the mistakes that characterized the "dot bombs" can be avoided.

HIGH TECH, BIO TECH, NANOTECH . . . .

There are precedents for predicting how the nanotech market will be divided between large and small. Looking at the way that the high-tech and biotech industries have developed, a trend emerges where smaller start-up companies tend to succeed with revolutionary technology and larger enterprises tend to succeed with evolutionary technology. For example, think of the great success stories of the high-tech market (primarily the information technology, Internet, and personal computer industries). They are primarily relatively new companies like Microsoft, Apple, Dell, Compaq, Yahoo!, eBay, and Cisco more than the old reliables such as IBM, HP, Siemens, Hitachi, and Sony. This is because home PCs and the Internet did not replace or evolve from older computers and older networks, but rather they were entirely new products that replaced nothing.

The biotech industry stands in sharp contrast to this. Although some of the methods of biotechnology, based on molecular biology, nucleic acid and protein chemistry, and molecular manipulation are new, the aim of most of this molecular wizardry has been continued progress in the established business of drug development. For this reason, the channels to market, expertise, and other assets of the grand old names like Merck, Glaxo, Pfizer, Lilly, Abbott, Baxter, and Pharmacia allowed these corporations to continue to dominate the industry. It is true that biotechnology has brought in several new big winners such as Genentech and Amgen, but because most of its products were not entirely new, most of its profits wound up in traditional groups.

Because nanotechnology is a method for improving many evolutionary and revolutionary technologies, ranging from something as straightforward as paint or glass or nonskid surfaces to such futuristic ideas as colorimetric nanosensors and neuro-electronic interfaces, it will have some aspects of both the biotech and the high-tech industries. Still, because many of the applications of nanoscience that are visible lie in the consumer, medical, agricultural, and energy arenas, it is probably true that most of the big hits will be owned by major industrial players who are already present in those markets. This does not mean that investments in start-ups, medium-sized companies, and growth firms centered on nanotechnology will not succeed. They certainly succeeded very well in the biotech business, mostly because these start-ups and young companies developed products, processes, know-how, and intellectual property. They became very attractive takeover candidates and were generally bought by big players. That is to the benefit of all involved. Investors and developers in small companies did very well economically; the products were developed rapidly and efficiently; and the development of the products into actual consumer marketplace items was handled by the strength, channels to market, and sophisticated distribution of the major players.

Shake-out and development patterns within nanotechnology are not yet clear, but it is the bet of the authors of this book that it will look more like biotech than like information technology or high-tech.

THE INVESTMENT LANDSCAPE

Although it is certainly possible to make money investing in large public companies with nanotechnology initiatives (and many large institutional investors and mutual funds will take this route), the most exciting way to invest in a rapidly developing economic sector is to invest in emerging companies. For that reason, despite our assertion that much of the action will be in the big firms, it's worth taking a closer look at what these emerging companies are likely to look like and what both companies and investors can learn from what has happened before. There has been a great hue and cry to "return to business fundamentals," but there are also great benefits in the easier access to capital and the start-up-friendly business environment and support mechanisms developed over the last few years.

Nanotechnology businesses are different from information technology and Internet businesses in several key ways. Nanotechnology is built on defensible intellectual property that can be patented and that may be difficult to replicate. The time required to duplicate a product or process is often measured in years, not months, and a competitor must take a substantially different approach or license the existing patent. In the Internet world, nothing prevented Barnes and Noble from developing a site just as good as Amazon.com and competing directly. The patentability of Web-based innovations is dubious at best as can be seen in Amazon's failed attempt to protect its patent on "one-click ordering" (it was deemed to be too obvious an innovation). It is now commonly accepted that "first mover advantage," the small headstart you get by being first to market with a new product, gives you a competitive edge for between 3 and 6 months, less time than it takes for many people to find out that you exist. After that period, someone will clone your product and probably incorporate lessons learned from watching the mistakes you made with the first version. If the new competitor is an established company, it will also use its customer base, media, brand, and advertising power to capture the market. This is less true of nanotechnology, provided the development is truly revolutionary. A new nanolithography process can be protected, and anyone who wants to use it must pay—even the biggest, most established companies. This can create a major barrier to entry and also lower the chances of companies getting crushed by the price pressure of near-perfect competition, as they have been in the telecommunications industry. It also opens an ethical conundrum. If companies are allowed to patent some basic natural processes such as genetic sequences, they can put a stranglehold on drug developments and other research. That has been an issue in the biotech world for some time.

With nanotechnology, there are few barriers to the market's adoption of most new products. For any Internet company to succeed, it needs customers on the Internet. During the emergence of dot coms, this meant that people had to be encouraged to learn to use computers, get online, and then get comfortable with making online transactions. This happened slowly, and it largely happened because of the billions of marketing dollars lavished on the market by the Internet start-ups, marketing dollars most of them will never recover. For nanotechnology, there is no such barrier. It does not take any new product adoption process to get someone to use a faster computer based on a nanochip or to take a nanopharmaceutical. Engineers who design products may have a learning curve associated with adopting new materials and processes, doctors may require more training, and machine shops and assembly lines may need to be refitted, but this learning curve is already common and will not affect demand for the new products once they become available.

So nanotechnology may have built-in competitive advantages and barriers to entry. It may be a new industry with little threshold to adoption. But despite all these ingredients that make an entrepreneur's or investor's mouth water, there are new difficulties as well. One of them is that the long product development cycle cuts both ways. Not only does it create a barrier to entry, but it also means that products might have a longer time to market. Investors who have gotten into the habit of expecting liquidity on their investments in two to three years may have to think again, especially if they are working with companies involved in very complex problems like nanoscale improvements in optics or computing. These are highly challenging and must also surmount systems integration and manufacturing issues to reach commercial success. Simpler nanoproducts such as sensors or smart materials may generate faster returns.

As with dot coms, the ideas for new companies will not necessarily come from people well suited to be entrepreneurs. They will come from scientists and engineers. These founders will need to be surrounded with strong management teams capable of managing money, people, and partnerships. They will need to have sales and product development teams as well as research teams, and they will be under increased scrutiny to keep their budgets tight and to develop profitable products. Venture capitalists and other equity investors may not easily understand the science behind new products, but they should understand the business and the people. In a dot com, it was relatively easy to eliminate troublesome founders or to override their decisions. In nanotechnology start-ups, this will be much more difficult. If the founders are producing intellectual property on which a company is based, they are irreplaceable, and firing them would be like killing the goose who lays the golden egg. While there are ways of preventing them from taking ideas to other enterprises, even this is difficult.

Nanotechnology start-ups will also have much to learn from biotechnology companies. After they invent a product or process, they can attempt to produce it, license it, or sell the rights (or company) outright. Many early players such as Applied Nanotech and Molecular Electronics are tending toward licensing and collecting royalties. This money could then be reinvested into related research, allowing the companies to diversify. Applied Nanotech, for example, already claims to have 68 patents and 86 more pending. The whole strategy here is to create cash flow with minimal overhead. If the cash flow is sufficient, it can represent an alternate liquidity strategy—companies won't need to go public or sell out for investors to get money out. This strategy can remove huge costs from ongoing operations since the reporting, auditing, registration, and other requirements of going public can sap so much of a company's time and resources. Many biotech concerns have still decided to go public and have done so successfully, but nanotech companies may not need to do this if there is sufficient research capital available.

Though it is one option, venture capital may not be the best route for early-stage companies trying to break into nanotechnology. Few venture funds have yet developed expertise in understanding nanotechnology so they may not be able to add anything other than money until the research approaches viability. Some alternatives for grants, seed capital, and first-stage investments come from government programs. These programs include Small Business Innovation Research (SBIR) and Small Business Technology Transfer (STTR), which can award amounts sufficient to get businesses off the ground and which have some of the resources of the National Nanotechnology Initiative behind them. These programs have been funding nanotechnology since at least 1996. Small Business Administration (SBA) loans and Small Business Investment Companies (SBICs), which are hybrids between government programs and private equity, are another avenue, as are a select group of venture capital firms (some of whom are mentioned in Appendix A) that have begun to gear up for nano.

It is interesting to note that government agencies are being embraced as nanotechnology partners in a way that they never were by high-tech companies. There are already nanotechnology special interest and advocacy groups such as the NanoBusiness Alliance, and politicians like Newt Gingrich ("Nano Newt" according to his Web page) and Joe Lieberman are becoming interested and involved. The U.S. government is committing significant resources to nanotechnology (though it is reducing its overall support of much basic scientific research) and taking a more active role in its development. In the past, high-tech (and, to a lesser extent, smaller biotech) companies more or less scorned Washington and relied almost entirely on private capital and resources. One reason for the difference is that nanotechnology is truly global: American companies dominate information technology and Internet markets, but many of the most exciting innovations in nanotechnology come from Europe and Asia. Without strong government assistance at the earliest stages, no country can maintain leadership in the science that can change everything.

Individuals wanting to make money on the nano scene might be best advised to wait for a little while. Very few nanotech companies are public, and judging ideas in nanotechnology takes a lot of background and expertise. If you had invested in market-indexed mutual funds for the last few years you would have done much better than many ill-informed angels and venture capitalists. The same is likely to be true of nanotechnology. In the words of Peter Lynch, "Know what you own, own what you know."

OTHER DOT COM LESSONS . . . .

Despite the bursting of the dot com bubble, it is still much easier for early-stage companies to get access to capital that it was at any time before 1998. While the system is far from perfect, there is still a huge amount of money subscribed to private equity investment firms that could be used for nanotechnology. But just as underinvestment can be a problem, overinvestment can also present difficulties. No matter how hot the technology and how urgent the need to bring it to market, there is still a limit to how much money a firm can absorb and apply usefully to its business. While a cash-starved firm will not be able to pay its bills, purchase lab equipment, or hire crucial scientists and technicians, an overfunded firm will often unnecessarily increase expenditures and overhead, driving up its underlying cost structure and reducing return on investment. This happens when firms get larger-than-needed offices, hire personnel before they are required, fail to do proper financial scrutiny on new products, and do not watch marketing and soft costs carefully.

Overinvestment is not just a problem with companies. When the International Monetary Fund or World Bank makes loans to countries or governments, they look at their absorption capacity. It is known that dollars above that level will be spent inefficiently or dissipated through corruption. Although not every company will be run like Boo.com (one of the worst offenders in the dot com world) and accounting companies have undoubtedly learned to be more rigorous post-Enron, absorption capacity is important to keep in mind. Over the last few years, the amounts of money raised in investment rounds have been determined more by how much an investor wanted a deal, what percentage of the company he wanted to own, and what valuation he wanted to place on it than by budgeting how much money a company actually needed to accomplish its goals and how its proposed valuation corresponded with the size of its market and merit of its plan. In early-stage nanotechnology companies, these pitfalls can be avoided by agreeing to project milestones at the onset, preparing detailed budgets, and keeping the business as small as possible and as large as necessary.

But creating project milestones may be difficult for investors in a field as technical as nanotechnology. To a great extent, this process as well as the refereeing of business ideas must be done by a qualified third party. Many professors at research institutions already act as consultants to large industrial firms, and investment companies may want to consider such formal relationships rather than the over-a-cup-of-coffee advisory relationships that are currently pervasive. This will result in a much more rigorous analysis of a new concept and a better notion of whether an idea is truly too good to be true, something that isn't always obvious in nanotechnology. While this consulting is expensive, it can save millions of investment dollars and also increase good deal-flow. Some nanotechnology institutes such as those at Northwestern are also setting up university-industry partnership programs, which may be perfect for this application. Smart investors bring expertise in business, risk management, and deal formation, but they shouldn't try to handle the technical side as well.

For all this, nanobusiness is still business, not some new revision of everything we know. Even though the technology is revolutionary and some of the support structures for start-ups have gotten stronger over the last few years, most of the old rules still apply. Businesses need a clear path to profitability. They need well-rounded and seasoned executive teams with the right people joining at the right time. They need to have the expectations of investors and executives kept in line with one another, and they need good working relationships at the top. Nanobusiness plans and models will not need to be complicated—most will be based on a biotech or independent R&D firm framework. The challenge is not engineering business to fit nanotechnology but rather engineering nanotechnology for business.

11 Nanotechnology and You

"As the discoveries of modern science create tremendous hope, they also lay vast ethical mine fields."

George W. Bush

In this chapter...

NANOTECHNOLOGY: HERE AND NOW

So what will life be like in the heyday of nanotechnology? How will its development actually impact people's day-to-day lives? The advances in creating ultra-fast computers, molecule-sized mechanical parts, and super-strong materials all sound grand, but what do they ultimately mean to me in my life?

These questions are coming from people who see the nanotechnology hype but aren't necessarily technophiles or long-term planners. Mentions of nanotechnology in the press have grown exponentially (see the nanotechnology hype index in Figure 5.4) as nano has become hot, but when will nanotechnology actually enter our lives and become not the next big thing, but the current one?

To some extent, the answer is "now." The original ideas behind nanotechnology go back some 20 years, but the first few nano-inspired inventions are just starting to hit store shelves. Even so, the first generation of nanogoods is still only a hint of what is to come.

Nanotechnology has been making its presence felt in industry for some time, and many applications are already standard. Because of the current national debate regarding energy policy and oil, a perfect example may be petroleum refining. Zeolites, the molecular sieves discussed in Chapter 6, are now used to extract as much as 40 percent more gasoline from a barrel of crude than the catalysts they replaced. This technique was first developed by Mobil and by some estimates saves approximately 400 million barrels of oil per year (around $12 billion) in the United States alone. Because this approach has been used for many years, don't expect it to drive down your pump prices any time soon, though it did when it was first developed. Even so, zeolites do show how significant (and how understated) the use of nanotechnology can be.

So what forms of nanotechnology are we most likely to see and touch? Perhaps first on the list of consumer nanogoods are smart materials such as coatings and laminates. Even though you may not put "coatings and laminates" on your grocery list, they are around you every day. In this case, they are thin layers of various materials that are engineered at the nanoscale to enhance other products in various ways. For example, the windows in new Audi A4 series cars are coated with glass laminates that block harmful ultraviolet radiation that can cause skin cancer. Additionally, the Institute for New Materials in Germany is manufacturing windows that contain a nanolayer of material that changes from clear to dark blue when a switch is thrown. This approach could be used as an alternative to window shades or window tinting, and some suppliers are now coating windows and other surfaces with ultra-hard scratch-resistant layers that may make car-keying a thing of the past.

In addition to windows, German and Japanese manufacturers such as Nanogate Technologies have started selling bathroom and kitchen tile that cannot get dirty since it is impossible for dirt and grit particles to cling to the coating in much the same way that food cannot stick to Teflon pans. These self-cleaning tiles can also be impregnated with biocidal (antimicrobial) nanoparticles. This prevents growth of rots and fungi that infest bathrooms and enhances overall sanitation. These tiles may put an end to the ever-unpleasant task of bathroom cleaning, a prospect many might consider sufficient cause to support all of nanotechnology.

But early support for nanotechnology did not come from bathroom cleaners; it came from groups like computer enthusiasts. This support emerged not just because computer users tend to be fans of new and emerging technology of all kinds, but also because nanotechnology offers so much to the world of computing. Even for those who don't particularly want a quantum computer on their desktops, a variety of very exciting products will soon be available. You won't be able to buy a Pentium-DNA for Christmas this year, but you will be able to start placing orders for some new kinds of computer displays.

Displays have been a focal point of computer engineering for last few years. Slowly, clunky TV-like cathode-ray tubes (CRTs) have been replaced by flat-panel liquid-crystal displays (LCDs). LCDs are more energy efficient, cause less eyestrain, and are more compact that CRTs. But the viewable area of LCDs is typically smaller than that of CRTs (few exceed 24 inches), their images are generally less bright, and they must be viewed directly, not from the side. Also, LCDs tend to refresh the images that they display slowly, which can cause animation and video to look sloppy. Enter light-emitting diode displays.

Everyone has seen LEDs. They are the bright pin lights often used on electronics as power and status indicators or as backlights. Using nanotechnology, these ultra-bright little lights can now be integrated on a panel densely enough to act as a display. By putting them in three-LED clusters (one red, one green, one blue) and mixing them while controlling intensity, any color can be created. Nano-based LED displays are the size of LCD displays but are even brighter than CRTs and allow for smooth, crisp animation. Also, like LCDs and unlike CRTs, LED displays don't require digital-to-analog conversion of images, a process that reduces picture fidelity. But LED displays have existed for some time; many signs already use them (such as the NASDAQ sign in Times Square, which contains some 19 million individual LEDs). What nanotechnology adds is the ability to make them small enough to put them into color-controllable clusters and pack them densely enough to create an image that is smooth to the human eye.

Nanotech is also taking a stab at improving the venerable CRTs. Using nanotubes to replace scanning electron guns, manufacturers like Samsung are shrinking these screens and reducing power consumption. It is even possible that these screens will be small, light, and efficient enough for use in laptop computers.

Another display technology, which one of its developers (the Irish company Ntera) is calling NanoChromics and another (E Ink) is calling electronic ink, is breathing new life into an idea from the 1990s that wasn't then feasible—digital paper. The idea behind the original digital paper was to create hand-held computers with ultra-sharp screens that could be held and read like a book. You could transfer digital files onto these computers so that you could store as many books and documents as you liked and eventually the world would become paperless. But there were a number of problems with these early displays: they were power-hungry, bulky, and never as easy to look at as a paper image (screen resolutions were still typically four times lower than print resolutions). Nanotechnology changes this: some new digital paper displays use the same chemicals that are used in paper to create a paper-like look, and picture elements, or pixels, are bistable so that once they are programmed to display a certain image they maintain that image without using additional power. Although people may still prefer the touch and feel of a book for many reasons, digital paper is very likely to take a substantial role in signage, since current technology for large format printing, error-correction, and shipping remains expensive.

Nanotechnology is not just present in fields that are traditionally high tech. Nano is now, literally, in fashion. Advances in molecular-scale composite materials have allowed companies like Nano-Tex to create next-generation cloth and clothing. Materials almost totally resistant to stains and materials that combine the comfort of cotton or natural fibers with the strength and durability of synthetics like nylon are already hitting the market in products from Eddie Bauer, Lee Jeans, and Nano-Tex parent, Burlington Industries. In addition to convenience and sharp looks, other nanotechnology-based fabrics incorporate the same sorts of biocidal agents present in the ever-clean bathroom tiles. These fabrics could be of great use in hospitals, where pathogens are common and patients are currently at significant risk from each other's infections.

Another industry in which new technology is frequently applied is sports equipment. Carbon fiber and graphite composites made their debuts in lightweight bikes and America's Cup sailboats. Fiberglass and plastics have been used for better football and hockey pads. And now nanotechnology is taking the field. Wilson's Double Core tennis balls use a nanocomposite clay to keep balls bouncing longer (they last two to four times as long according to Wilson and are now the official balls of the Davis Cup), and Babolat has introduced super-strong nanotubes into its tennis racket line for improved torsion and flex resistance. Nanotubes are sure to see wider adoption in sports equipment as their prices come down (nanotube-enhanced golf clubs are on the way), but for now we'll mostly see them in the hands of the pros while we are in the stands.

A last area where nanotechnology truly shines is in medicine. While medicine has not traditionally been considered a consumer market, nanotechnology may be changing that. Home pregnancy tests have already seen improvements in ease of application, speed of results, and overall accuracy since they have started employing nanoparticles, and other home tests are becoming feasible. Some scientists hope to see tests for everything from anthrax to AIDS made simple enough for self-application through the use of nanotechnology, and goods like braces and prosthetics are already targets of early nanotechnology ventures.

All in all, while much of the promise of nanotechnology remains in the future, it is already slipping into our lives through our houses, our computers, our games, and even our bodies. The age of nanotechnology is truly upon us.

NANO ETHICS: LOOKING BEYOND THE PROMISE OF NANOTECHNOLOGY . . . .

As nanotechnology steps onto center stage and the bounty it promises begins to become reality, it will raise several issues of ethics, public policy, law, and social responsibility. Most of the questions aren't new, but nanotechnology increases the urgency and importance of addressing them.

Perhaps the most important questions in the short term will be issues surrounding patents. Biotechnology and pharmaceuticals are two segments of industry that stand to gain a great deal from nanotechnology because treatments or even cures for many of the world's most virulent illnesses may be possible through nanotechnology. Under current laws in many countries, developers and research companies can patent drugs as well as genetic patterns and synthesis techniques. Researchers are afforded this patent protection for the same reasons that other firms in other industries are awarded patents—to encourage them to innovate and to allow them to recover the costs of research, development, and testing of their products.

This argument is not without merit. Pharmaceutical research is very expensive and risky. Few drugs that start the development process are deemed effective, pass the rigorous regulatory and approval processes, and ultimately make it into hospitals and pharmacies. Generic drug makers can cut a huge amount of cost by avoiding this and going directly to market, creating unfair competition if a patent does not block them. But many doctors, governments, and social activist organizations argue that such patents not only allow reasonable recovery of drug-development costs but also keep discoveries secret so that they can't be used to spur further development, and so are used for profiteering by the big multinational pharmaceutical companies. One such case involves the multidrug protease-inhibitor "cocktails" used to treat AIDS. In 2001, it cost around $10,000 to purchase the drugs needed annually for the treatment of one person in the United States. This pricing put the drugs completely out of the price range of individuals living in countries in Africa and South America where average incomes are at most a couple of thousand dollars per year. Generic drug manufacturers in other counties (such as Cipla in India, now approved by the World Health Organization) produced the drugs for a fraction of the cost, making it possible for Brazil to implement one of the world's most successful anti-AIDS programs. While this might seem like a victory—obviously everyone wishes AIDS could be reduced or treated—the patent-holders objected and tried to block generics makers. They claimed that they could not continue to research new drugs such as a true cure for the disease if they were not allowed to recover their costs.

A similar case happened in the wake of September 11, when governments sought to purchase large quantities of ciprofloxacin, an antibiotic useful in the treatment of anthrax. Patent-holder Bayer significantly reduced the price it offered the U.S. government, but generics manufacturers such as Apotex of Canada still offered the drug for much less, though Bayer sought to block them. The question that emerges from both of these cases is when, if ever, public health and benefit should be prioritized above patents and other restrictions. If nanotechnology starts to deliver treatments for cancer or even all-purpose frameworks for antiviruses, these questions will become paramount.

Beyond the patent question, nanotechnology will have societal and geopolitical implications. The first industrial revolution created significant divisions between the industrialized economies and what have been variously called underdeveloped, less-developed, or Third World countries. This is in part because industrialized countries can produce goods less expensively and are much wealthier. There is a huge barrier to entry for industrialization—countries must invest in education, infrastructure, political reform, law enforcement, and modern manufacturing facilities. The capital to do this tends to come from nations that have already industrialized; as a result, emerging economies can find themselves beholden to the already rich countries. This has been the argument behind confrontations ranging from the Cuban revolution to the antiglobalization protests in Seattle. Nanotechnology could potentially increase that divide between rich and poor nations. It could even eliminate the advantages of countries rich in natural resources such as oil since it may make the cheap synthesis of materials and solar energy conversion possible. Alternatively, if nanotechnology fabrication turns out to be inexpensive and easier to disseminate than current industrial technology (as is the case with information technology, which has led to its success in places like India), it could reduce the gap between rich and poor or at least make it easier for people around the world to get their basic needs met.

On a more technical front, nanotechnology has already excited many with its defense and weapons applications. These could include better armor and improved battlefield communication. But each application has a flip side worth examining. For example, superstrong nanomaterials such as nanotubes are predominantly carbon-based. This means that they cannot be detected using metal-detectors or chemical "sniffers." The only way to catch a person smuggling such a weapon onto an airplane or into a building would be by exhaustive personal searches. Other even more sinister applications from the darker pages of science fiction could conceivably be possible. For example, some organizations might seek to create viruses that only target people with specific genetic characteristics or even to customize a virus for a specific person. The degree to which such biological weapons would be able to target specific groups is not clear, but it is certain that genocidal leaders such as Saddam Hussein would be eager to find out. And with nanotechnology, unlike current biological weapons, there is no reason to believe that a researcher needs access to restricted existing viral strains or controlled materials. Potentially they could create such viruses essentially from scratch since none of the raw materials are as rare as enriched uranium.

These nightmare scenarios seem somewhat far-fetched and are certainly unlikely, but the whole issue of working at the scale that is the basis of life creates intrinsic ethical problems. There is already much consternation about the ethics of human and animal cloning. Both it and the debate over stem cells, which essentially asks whether pre-embryonic life can be sacrificed in order to create treatments that can prolong and improve the lives of people suffering from Alzheimer's, Parkinson's, juvenile diabetes, or other degenerative diseases, caused President Bush to take counsel with religious leaders and ethicists as well as scientific authorities to come up with a policy. It is rare for a president to become directly involved in determining the direction of a specific area of scientific research. The results, statements that use terms like "sacred," which are not often seen in scientific papers, show that the debate on nano- and biotechnology is much more touchy than research that involves finding new elements, new planets, or even new ways of causing nuclear reactions. And these are just the stirrings of the great debates yet to come. Advanced DNA analysis technology made by companies such as Affymetrix (which hosts ethics discussions on its Web site) and Agilent not only assists in finding genetic fingerprints of diseases but also potentially allows for the screening of healthy fetuses or the tracking of an individual's predisposition to disease. Such information would clearly be of interest to insurance companies and employers. This could rekindle the abortion debate and also discussions over an individual's right to privacy, this time over his or her own genetic data. And soon genetic engineering, already contentious with plants and crops, will become a possibility for humans. The fundamental question of to what extent we should tinker with ourselves must be addressed.

Even more futurist concerns also arise from nanotechnology. Arguably, forms of nanocomputation that we've discussed (such as quantum computing, DNA computing, and nanoelectonic computing) may help unlock true artificial intelligence. If this occurs, how should artificial intelligence be treated? What rights and privileges should it have? What if it should become self-replicating? Also, if interfaces between humans and computers improve to the point where they are hard to differentiate, what will that mean for human civilization? How will we treat these cyborgs? Would this truly be the next stage in the evolution of the human race? Might it allow for the arbitrary extension of life through artificial organs or bodies? Even if nanocomputation fails to produce machines that think, one of its stated goals is to break codes. If this reaches fruition, all common forms of digital cryptography from the sort that protects e-commerce to the kind that protects nuclear secrets could be compromised. The implications for national security and for personal privacy cannot be overstressed.

These are all heady questions, and it is time to begin to tackle them somewhere other than on TV and in the pages of science fiction books. The whole ethical debate over nanotechnology is one of the most important reasons for the public to know what nano is and what it could mean. Nanotechnology is already, by its very nature, a multi-

A Some Good Nano Resources

There is a time for some things, and a time for all things; a time for great things, and a time for small things.

Miguel de Cervantes Saavedra

FREE NEWS AND INFORMATION ON THE WEB

Many sites on the Web have information about nanotechnology and nanoscience. A few of them are regularly updated or have quality reporting. Many are sensationalist and need to be taken with a grain of salt. A list of some of the sites that we've found most useful follow.

NanotechBook: www.nanotechbook.com. This is the official Website for this book. On the site you will find copious links to other good sites as well as references for the book and some additional information and online discussions about nanotechnology.

Small Times: www.smalltimes.com. This is a good news-compilation site that concentrates on microelectromechanical engineering, microsystems, and nanotechnologies. It is also home of the Small Times Stock Index.

Scientific American Nanotechnology: www.sciam.com/ nanotech. This site is an excellent resource for breaking scientific news about nano.

The National Nanotechnology Initiative: www.nano.gov. This overview of the federal government program for nanotechnology links to some good educational resources.

The Nanotechnology Bulletin: www.nanotechbulletin.com. This site contains interviews with movers and shakers in nanotechnology. If you want to hear what experts inside the industry are saying, this is a good resource.

The NanoBusiness Alliance: www.nanobusiness.com. This site is a good source for those interested in the business and political side of the industry. It is also a good place to find out about events and conferences related to nanotechnology.

Nanotech Now: nanotech-now.com. This site is a portal that links to other good sites about nanotechnology.

Nanotech Planet: www.nanotechplanet.com. This is another good headline and portal site, although it is liable to mix nano with micro and other semirelated news. It also has a nanotech stock portfolio and an associated convention.

VENTURE CAPITAL INTERESTED IN NANO . . . . .

Nanotechnology has begun getting serious interest from venture capital firms, and some new firms are being formed just to get in on the action. Following is a brief list of a few venture capital firms that solicit nanotechnology proposals, have invested in nanotechnology companies, or have been active in promoting nanotechnology. This listing is not meant to be a recommendation or endorsement of any particular firm. There are many more, and the list is sure to continue growing as more products become practical.

AGTC Funds: www.agtcfunds.com

Angstrom Partners: www.angstrompartners.com

ARCH Venture Partners: www.archventure.com

Ardesta: www.ardesta.com

Ben Franklin Technology Partners: www.sep.benfranklin. org

Bessemer Venture Partners: www.bvp.com

Capital Stage Nano: www.capitalstagenano.com/en

CW Group: eee.cwventures.com

Draper Fisher Jurvetson: www.drapervc.com

Evolution Capital: www.evolution-capital.com
Galway Partners: www.galway.com
Harris & Harris Group: www.hhgp.com
Illinois Partners: www.illinoispartners.com
Lux Capital: www.luxcapital.com
McGovern Capital: www.mcgoverncapital.com
Morgenthaler Ventures: www.morgenthaler.com
Polaris Venture Partners: www.polarisventures.com
Portage Ventures: www.portageventures.com
Sevin Rosen Funds: www.sevinrosen.com
Tribal Weave: www.tribalweave.com
Venrock Associates: www.venrock.com

Glossary

Note: These definitions are intended to be convenient rather than complete. As such, they are restricted to the particular senses in which the word is used in this text and are simplified for clarity.

Amorphous Computing: See *Swarm Computing*.

Amorphous Polymers: Polymers that do not form a crystalline structure, but are disordered in the solid phase (e.g., polystyrene in cups).

Analytes: Chemical species, in solution or in the gas, whose presence and concentration are to be analyzed or sensed.

Antibody: Proteins produced by the immune system to neutralize or destroy antigens.

Antigen: A foreign substance that, when introduced into the body, becomes harmful because of immune response.

Artificial Photosynthesis: Application of molecular or solid state structures to imitate natural photosynthesis by using light to cause electrical current to flow. Artificial photosynthesis is one approach to photovoltaics.

Atomic Force Microscope: A scanning probe instrument that measures the force acting on a tip as it either slides along a surface or moves perpendicularly to a surface.

Bio-availability: The term used to describe the local availability, within a large biological entity like a human body, of a particular drug or therapy molecule.

Bio-inert Materials: Materials that do not react with the biological environment. Normally, bio-inert materials are not rejected by the human immune system.

Biosensor: A sensor structure that targets biological analytes or a sensor based on the use of biological molecules.

Bistable: A system that has two stable resting states, like a coin that can rest either on heads or on tails.

Bottom-Up Nanofabrication: The building of nanostructures starting with small components such as atoms or molecules.

CAD: Computer Aided Design, referring to the use of computational algorithms and tools to design switches, computers, memories, and other technology devices.

Catalyst: A substance that causes a chemical reaction to proceed more rapidly; for example, the crystallization of sugar on a wooden stick occurs more rapidly than it does in simple liquid water, so the stick acts as a catalyst.

Ceramics: Hard, refractory materials often based on oxides.

Chaperones: Small proteins used within cells to carry metal ion species from one place to another (this facilitating is quite opposed to the usual meaning of chaperones, whose business it is not to facilitate).

Cholesterol: A large molecule very widespread in biology (e.g., it provides much of the mass in human livers and brains); it is hydrophobic.

CMOS: Complementary Metal Oxide Semiconductor, referring to a technology commonly used to make contemporary silicon-based microchips.

Colorimetric Sensors: Sensors that work by changing their colors when the analyte appears. A simple example is litmus paper.

Coulomb's Law: The fundamental law of electrical interaction: the force between two charges is proportional to the magnitude of each charge and inversely proportional to the square of the distance between the charges.

Cross-Linked Polymers: Structures consisting of linear polymer strands with chemical bonds between them that act to link one strand to the next.

Crystal Growth: The formation of crystals by growth from solution, a kind of self-assembly. Examples include the formation of snowflakes in the atmosphere and the formation of rock candy from sugar solutions, showing that sweet and beautiful are both forms of nanotechnology.

Decoherence: The loss of controlled entanglement. For example, when two entangled pieces of information (in quantum computing) become separated by the loss of relative phase, they are said to decohere.

Dip-Pen Nanolithography: Method of fabrication that uses a scanning probe tip to act essentially as a fountain pen to draw nanostructures on surfaces using arbitrary molecular structures as inks.

DNA Computing: The use of DNA hybridization and replication processes to solve computational problems.

DNA Molecule Therapy: Therapeutic scheme in which DNA molecules are introduced into the cell, where they can combine with, for example, pathogenic DNAs within infected cells.

Electrochemistry: The science that combines chemistry with the flow of electrical current. Examples of electrochemical processes include silver plating and aluminum manufacturing.

Electron: The subatomic particle with one negative charge and a mass that is roughly 1/2000 the mass of a proton.

Electron Beam Lithography: Method of fabrication that uses electron beams to form structures on surfaces. It is widely used to make extended large nanostructures.

Electron Microscopy: The measurement of structures of solids and surfaces using electrons rather than light to see small features (down to the nanoscale).

Electroosmosis: A method for moving liquids using electrical fields. Electroosmosis moves a sample at a constant rate, so it is used when a sample should not separate into components. Compare to *electrophoresis.*

Electrophoresis: A method of using external electric fields on electrically charged species to move either particles or liquids. Electrophoresis moves samples at a rate proportional to inverse mass of the components of the sample and is therefore often used for separations. Compare to *electroosmosis.*

Encapsulated Nanomaterials: Structures in which a nanomaterial is enclosed in an outer covering or coat.

Entanglement: In quantum computing, the process of combining two separate pieces of information so that they can be treated as a single entity.

Enzyme: A protein catalyst used to facilitate chemical reactions in biological species.

Fab: Abbreviation for fabrication, used to refer to a plant that manufactures semiconductor devices (microchips).

Field-Effect Transistor: The most common sort of transistor used in semiconductor chips. It employs a gate to control (turn on and off) electrical current.

Finite State Machine: A device that operates by switching among a series of states according to a set of rules (see *Transition Rules*). An example of a Finite State Machine is an elevator controller. Each position of the elevator (first floor, second floor, etc.) represents a state and how the elevator reacts to users pressing the call buttons is encoded in the Transition Rules.

Gene Therapy: A specific sort of DNA therapy.

Giant Magneto Resistance: A phenomenon in which the resistance of a substance is changed very strongly by application of a magnetic field. It is used as a read-out mechanism in current magnetic computer memories.

Graetzel Cell: A photovoltaic cell, first developed by Michael Graetzel in Switzerland, that uses nanoscale titanium dioxide and an organic dye to obtain electrical current from incident light.

Head Groups: Part of the structure of some long molecules wherein one end of the molecule can be called a head group and the other end a tail group. In soap, the hydrophilic head groups are soluble in water, while the hydrophobic tails cause solubility in oils and greases.

Histamine: A small molecule that, although always present in the body, is increased in concentration by the presence of antigens and antibodies. Histamines often cause allergic responses.

Hybridization: In DNA science, the formation of a second strand from the first, by complementary binding of the G with C or A with T.

Hydrogen Bond: A specific sort of weak bond between molecules caused by a hydrogen atom bridging between (say) an oxygen to which it is covalently bound and another oxygen that interacts more weakly with it by electrical forces. Hydrogen bonds are crucial for the structure of water, proteins, and DNA.

Hydrophilic: Water loving, refers to materials or molecular structures that interact strongly with and are soluble in water (e.g., ethyl alcohol).

Hydrophobic: Water hating, refers to materials that do not dissolve in water (e.g., salad oil).

Inhomogeneous Structures: Materials that are not the same throughout. Macroscopic examples are a western omelet and reinforced concrete.

Insulators: Materials that do not permit electricity to flow (e.g., the rubber lining on extension cords).

Ion: An atom or molecule that is electrically charged.

Light-Emitting Diodes: Structures in which an electron and a hole combine to form an excited state that subsequently emits light; these devices permit direct transformation of electricity into light.

Light-Harvesting Complex: The part of the photosynthetic apparatus that actually captures and stores the light energy (as molecular excited states) before passing it on to other structures within the photosynthetic apparatus.

Liposome: A microscale or nanoscale artificial globule consisting of layers of lipid or phospholipid enclosing an aqueous core. It can be used both as a model for membranes and as a delivery vehicle for particular molecules or biological structures.

Lithograph: Originally meant a pattern or structure formed on paper using ink or painted stones to form the image.

Lithography: The formation of structures of any size (including nanoscale), generally by transferring the pattern from one structure to another.

Logic Gates: The fundamental logic structures that, combined, lead to digital computing. The three most common logic gates are AND, OR, and NOT. Logic gates were first discussed by George Boole.

Luminescent Tags: Molecules or nanostructures that luminescence (emit light) when illuminated and that are used to identify structures to which they are bound.

Macromolecule: Another word for polymer; it refers to single molecules consisting of many (thousands or more) atoms.

Magnetic Force Microscope: A scanning probe microscope in which a magnetic force causes the tip to move. This motion allows the user to measure the magnetic force..

Magnetic Resonance Imaging: A form of magnetic resonance spectroscopy that indicates the presence of particular atomic nuclei. Used to image sections of the body or particular biological structures.

MEMS: Microelectromechanical Systems, referring to structures at the micron scale that transduce signals between electronic and mechanical forms.

Metastasis: The process by which certain cancers can spread from one organ or structure within the body to another.

Microfluidics: The process of moving liquids or fluids along a channel whose characteristic cross-sectional dimension is microns.

Micro Imprint Lithography: Lithographic method for making small structures (originally at the micron, now at the large nanoscale level) using a sort of ink pad, usually made of a plastic material.

Microtubules: Extended rigid linear structures found in cells; they are used by molecular motors such as actin to move cargos of molecules or other structures within the cell.

Mirroring: A method for keeping stored data consistent among two or more digital storage media, such as computer hard drives.

Molecular Conductors: Molecules that can conduct electrical current.

Molecular Electronics: Electronics that depend on or use the molecular organization of space.

Molecular Motors: Complex nanostructures (sometimes slightly larger than nanostructures) that work to transform chemical energy to mechanical motion within biological structures.

Molecular Recognition: Fundamental self-assembly scheme wherein one molecule has the ability to bind in a specific way to another molecule or a surface.

Monomers: Small molecules that bind together into longer structures to form polymers.

Nanocomposites: Composite structures whose characteristic dimensions are found at the nanoscale. An example is the suspension of carbon nanotubes in a soft plastic host.

Nanodots: Nanoparticles that consist of homogeneous material, especially those that are almost spherical or cubical in shape.

Nanofabrication: The manufacture or preparation of nanostructures.

Nanofiltration: The filtering of particles of nanosphere dimensions.

Nanofluidics: The process of moving liquids or fluids along a channel whose characteristic cross-sectional dimension is nanometers.

Nanorods: Nanostructures that are shaped like long sticks or dowels, with a diameter in the nanoscale and a length very much longer.

Nanoscale: Refers to phenomena that occur on the length scale between 1 and 100 nanometers.

Nanoscale Biostructure: A biological structure whose characteristic properties change on the nanometer length scale (e.g., a cell wall).

Nanoscale Synthesis: Another word for nanofabrication, referring to manufacture of structures at the nanoscale.

Nanoscience: A discipline in which the authors of this book work, involving scientific understanding and investigation of nanoscale phenomena.

Nanosphere Lift-Off Lithography: A nanofabrication method in which small spheres of nanoscale dimension are used to form a pattern on a surface, which then acts as a mask to block some areas of the surface during subsequent deposition of a nanomaterial from the vapor phase. It is a nanoscale version of letter stencils that are used to spray paint signs.

Nanostructures: Structures whose characteristic variation in design length is at the nanoscale.

Nanotechnology: The application of nanoscience in technological devices.

Nanotubes: Almost always carbon nanotubes, referring to the wires of pure carbon that look like rolled sheets of graphite or like carbon soda straws.

Nanowires: Another term for nanorods, especially nanorods that can conduct electricity.

Neuro-Electronic Interface: A structure that permits transduction of signals between nerve fibers and external computational resources.

Neurotransmitters: Small organic molecules that carry signals and information from one part of the brain (neuron) to another.

Neutron: A subatomic particle with no electrical charge and a mass slightly larger than a proton; it can be thought of as a combination of a proton and an electron.

Ohm's Law: Fundamental law of electrical charge flow in macroscopic circuits stating that the current is equal to the voltage divided by the resistance.

Oligonucleotides: Small subunits of DNA consisting of a few bases on each of the hybridized strands. "Oligo" means few.

Optics: The science of light and its propagation and interaction with matter.

Pervasive Computing: A futuristic scenario in which essentially all the functional structures of life ranging from door locks to kitchen stoves to raincoats to humidifiers are computer controlled and in mutual contact.

Photodynamic Therapy: A remediation scheme for several diseases, including cancer. It depends upon the use of molecular or quantum dot structures to transform light energy either into heat or into highly reactive excited oxygen molecules that subsequently attack the tumor tissue.

Photorefractive Polymers: Polymeric materials that exhibit both charge motion and nonlinear optical response so that patterns of information can be written and read using them.

Photosensor: Usually, a device for measuring the presence and frequency of light. The most common photosensors are macroscopic, work by emission of electrons from photoexcited metals, and are used, for example, in the emergency openers for elevator doors.

Photosynthesis: The process by which plants and bacteria transform light energy into chemical energy, molecule synthesis, or proton gradients. It is the fundamental means by which nearly all energy sources are powered by the sun.

Photovoltaics: Artificial systems that transform light energy into electrical current; they can be based either on semiconductor structures or on molecular complexes.

Pipelining: A microprocessor design approach that involves breaking up each individual processor instruction into smaller subinstructions. Each subinstruction requires only part of the processor, so multiple instructions can be processed at the same time.

Polymerization: The process of making polymers from monomers, thereby making very large macromolecules from small molecular precursors.

Polymers: Extended molecules made by bonding together subunits called monomers. Examples include polystyrene and polyethylene, as well as DNA.

Polysaccharides: Polymers whose subunits are sugars.

Prestin: Molecular motor structure found in the inner ear and important in transducing sound into neural signals.

Protein Engineering: The manufacture and manipulation of proteins by synthetic chemical routes.

Proteins: Biological macromolecules assembled from amino acid units. They are the functional structures in biology.

Proton: A subatomic particle with a positive charge of one unit and mass slightly smaller than a hydrogen atom. The number of protons in a given nucleus determines which element the atom is.

Quantum Computing: A computing scheme that depends upon the wave-like properties of matter and that works in a way that is fundamentally different from digital computing.

Quantum Dots: Nanostructures of roughly spherical or cubic shape that are small enough to exhibit characteristically quantum behavior in optical or electrical processes.

Quantum Mechanics: A description of the mechanical behavior of atomic and subatomic particles such as electrons and protons. Quantum mechanics is a generalization of classical mechanics, which describes basketballs and horseshoes.

Qubit: The smallest unit of information in quantum computing.

Quinones: Small organic molecules containing double bonds between carbons and oxygens. They are important as intermediate acceptor species in photosynthetic structures.

Rodcoils: Medium-size molecules containing hundreds to thousands of atoms that are arranged with a stiff tail and a soft, hydrophobic, space-filling bulbous head. They self-assemble into extended round and cylindrical structures.

Scanning Probe Instruments: Tools for both measuring and preparing nanostructures on surfaces; they work using the interactions between a scanning tip structure and the nanostructure on the surface, which can be either manipulated or measured.

Scanning Tunneling Microscope: The first of the scanning probe instruments, invented by Binnig and Rohrer. It works at the scale of nanostructures and measures electrons tunneling between a scanning tip and a conducting surface.

Self-Healing Structures: A form of smart material in which the structure responds to a physical stress, break, or fracture by repairing itself back to the original structure.

Siderophores: Small molecules containing oxygen, nitrogen, sulfur, or phosphorous atoms that can bind to (capture) particular metal ions.

Species: In chemistry refers to a particular atom, ion, or structure.

Spectroscopy: The science of the interaction of radiation with matter.

Suicide Inhibitors: Synthetic molecules that, upon reacting with an enzyme, produce a product that binds to the enzyme and therefore causes the enzyme not to function (to commit functional suicide).

Swarm Computing: An alternative computer architecture based upon a very large number (a swarm) of very simple devices instead of a small number of megalithic microprocessors. Each device can perform only a few elementary tasks and each can fail without disrupting the system. Swarm algorithms are complex and are designed to take advantage of both limited strength and the possible failure of a given computational element.

Tail Group: See *Head Groups*.

Top-Down Nanofabrication: The process of making nanostructures starting with the largest structures and taking parts away. It is analogous to classical sculpture and to CMOS chip fabricating. To make David, it is said that Michelangelo started with a block of marble and took away everything that wasn't David.

Transduction: Process of changing energy or signals from one form to another.

Transition Rules: A scheme of instructions that tells a finite state machine (a form of computing device) to move from one particular state (e.g.. on or off) to another. An example of a transition rule is "Turn on when the power switch is thrown."

Ultrafiltration: The filtering of very small (micron-scale) particles. It is not a precise usage, and sometimes people refer to nanofiltration as a form of ultrafiltration.

Zeolite: A framework ceramic material built of aluminum oxide and silicon oxide, with other possible additions. Used for water softening and for several catalytic structures, they comprise a very elegant set of nanoscale materials.

Index

A

B

C

D

E

F

G

H

I

K–L

M

O

P

Q

R

S

T

U–Z

About the Authors

Professor Mark Ratner is Morrison Professor of Chemistry and Associate Director of the Institute of Nanotechnology and Nanofabrication at Northwestern University. His lifelong work in molecular electronics, a field he is credited with creating in 1974, led to his receiving the 2001 Feynman Prize in Nanotechnology and becoming a member of both the National Academy of Sciences and the American Academy of Arts and Sciences. He has published two advanced textbooks on chemistry, nanotechnology, and related subjects and more than four hundred scientific papers. Professor Ratner has been a named lecturer on molecular electronics and nanotechnology at institutions across the world, but he concentrates his efforts on Northwestern University (where he received his Ph.D., served as Associate Dean of the College of Arts and Sciences and Chair of the Chemistry Department and received the Distinguished Teaching Award).

Professor Ratner holds a B.A. from Harvard University and is a former Director of Electrochemical Industries. He has held fellowships from the A. P. Sloan Foundation, the Advanced Study Institute at Hebrew University, the American Physical Society, and AAAS. He also serves as vice-chairman of ideapoint, a Midwest regional venture capital and research company specializing in nanotechnology. When not doing research, Mark can be found fly fishing knee deep in icy water or canoeing on a lake in Maine.

Dan Ratner is a veteran of high-tech startups. He was a co-founder, and he currently serves as vice president and CTO of Driveitaway.com, a Web-based business specializing in dealer-to-consumer car auctions. Prior to Driveitaway.com, Mr. Ratner was the co-founder and CTO of ISP Wired Business, a pioneer in the distribution of DSL-based Internet access. He began his start-up career as the founder and CEO of Snapdragon Technologies, a technology and business consulting firm specializing in information systems and strategies for clients nationwide. In July 2001 Mr. Ratner was selected by *PhillyTech* magazine as one of the "Thirty Under 30" entrepreneurs to watch in the Philadelphia area. Prior to his involvement with startups, he was an electrical engineer at Zeller Research Ltd.

Mr. Ratner holds a B.A. in Engineering and Economics from Brown University and is a Visiting Scholar at Northwestern University. He sits on the Board of Directors of Sittercity Inc., the Boards of Advisors of First Colonial National Bank and RMS Investment Corporation, and serves as a mentor with the Brown University Entrepreneurship program. Recently, he lectured on nanotechnology and business at the Kellogg School at Northwestern University. When not ruminating on a new venture, Dan can be found ballroom dancing or collecting wine.

What the experts are saying:

"*Nanotechnology and Homeland Security* provides the reader with the most important weapon of all—knowledge. It is as much a blow against ignorance and hype as it is a primer for how real nanotechnology should contribute to our future security. Mark and Dan Ratner confront the utopians and the alarmists by debunking both 'molecular assemblers' and 'gray goo'. This book is informative, thought-provoking and very readable."

— R. Stanley Williams
HP Senior Fellow, Hewlett-Packard Labs

"The book is a clear overview of the two subjects of nanotechnology and countering terrorism, but its special strength is the thoughtful way it weaves these two subjects together."

— R. Stephen Berry
Department of Chemistry and the James Franck Institute, The University of Chicago, Chicago, Illinois

"This book does an excellent job introducing the field of nanotechnology to the layperson by showing its promise for security and defense—perhaps the most relevant sectors of society demanding advances that only nanotechnology can provide."

— Josh Wolfe
Managing Partner, Lux Capital & Editor, Forbes/Wolfe Nanotech Report

"U.S. policy-makers and -shapers: READ THIS BOOK! Then get to work."

— Rocky Rawstern
Editor Nanotech-Now.com

"This book identifies many of the issues that need to be examined, and to be dealt with, if nanotechnology is to become a fully mature, fully productive asset to our nation and to the world."

— James Murday
Chief Scientist, Office of Naval Research

"Mark and Dan Ratner have ably illustrated some of the roles that nanotechnology can play in our future, including how it could enhance national security, make soldiers more effective on the battlefield, or even help prevent attacks on our homeland. As a member of Congress who is active in advancing the development of nanotechnology, I encourage other policymakers, educators, and social visionaries to become cognizant of tomorrow's possibilities."

— U.S. Representative Mike Honda
Member, House of Representatives Committee on Science and Co-Author of the *Nanotechnology Research and Development Act of 2003*

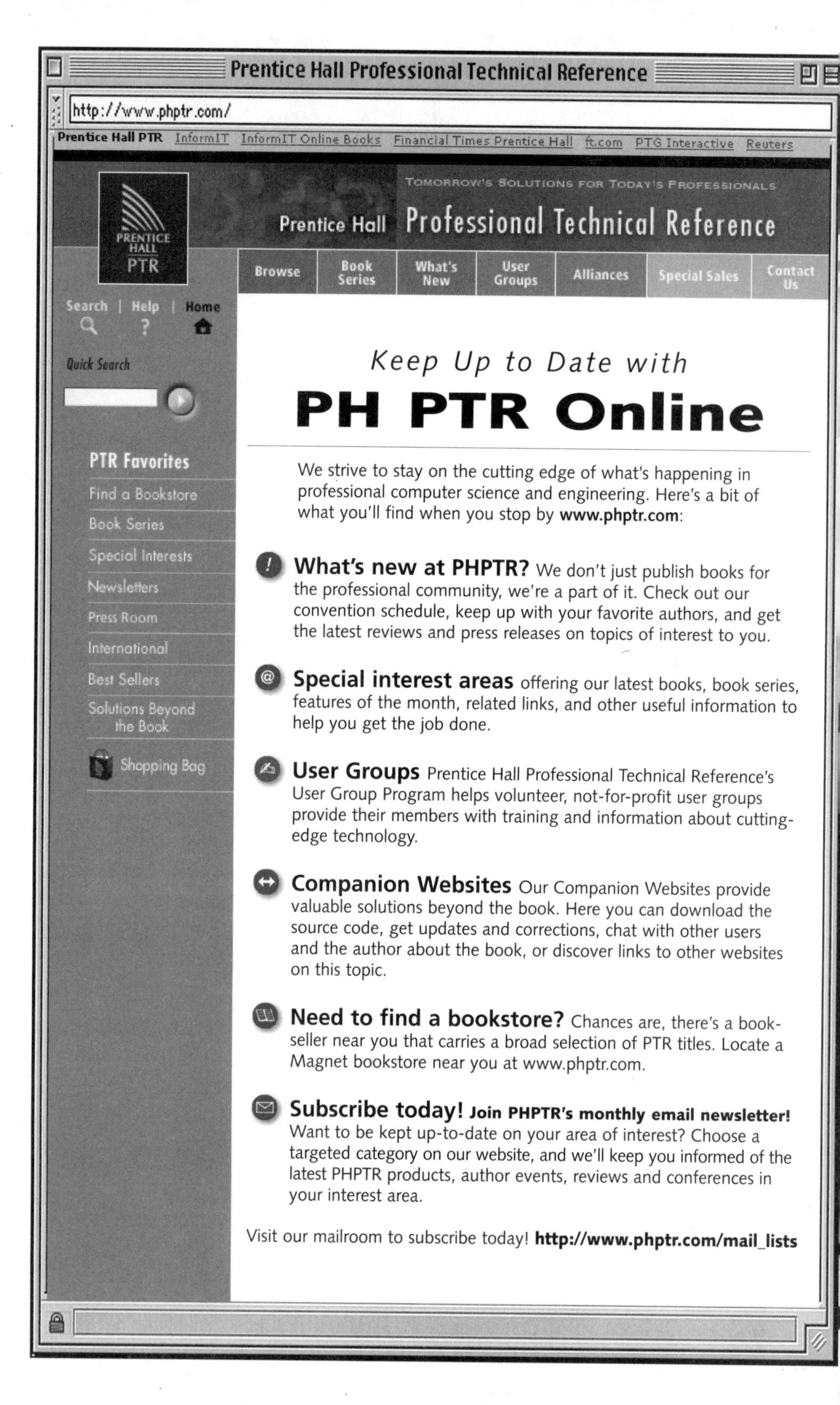

Prentice Hall Professional Technical Reference
http://www.phptr.com/
Prentice Hall PTR
InformIT
InformIT Online Books
Financial Times Prentice Hall
ft.com
PTG Interactive
Reuters
PRENTICE HALL PTR
TOMORROW'S SOLUTIONS FOR TODAY'S PROFESSIONALS
Prentice Hall Professional Technical Reference
Browse
Book Series
What's New
User Groups
Alliances
Special Sales
Contact Us
Search | Help | Home
Quick Search
PTR Favorites
Find a Bookstore
Book Series
Special Interests
Newsletters
Press Room
International
Best Sellers
Solutions Beyond the Book
Shopping Bag
Keep Up to Date with
PH PTR Online
We strive to stay on the cutting edge of what's happening in professional computer science and engineering. Here's a bit of what you'll find when you stop by www.phptr.com:
What's new at PHPTR? We don't just publish books for the professional community, we're a part of it. Check out our convention schedule, keep up with your favorite authors, and get the latest reviews and press releases on topics of interest to you.
Special interest areas offering our latest books, book series, features of the month, related links, and other useful information to help you get the job done.
User Groups Prentice Hall Professional Technical Reference's User Group Program helps volunteer, not-for-profit user groups provide their members with training and information about cutting-edge technology.
Companion Websites Our Companion Websites provide valuable solutions beyond the book. Here you can download the source code, get updates and corrections, chat with other users and the author about the book, or discover links to other websites on this topic.
Need to find a bookstore? Chances are, there's a bookseller near you that carries a broad selection of PTR titles. Locate a Magnet bookstore near you at www.phptr.com.
Subscribe today! Join PHPTR's monthly email newsletter!
Want to be kept up-to-date on your area of interest? Choose a targeted category on our website, and we'll keep you informed of the latest PHPTR products, author events, reviews and conferences in your interest area.
Visit our mailroom to subscribe today! http://www.phptr.com/mail_lists